YOUR KNOWLEDGE HAS VALUE

- We will publish your bachelor's and
  master's thesis, essays and papers

- Your own eBook and book -
  sold worldwide in all relevant shops

- Earn money with each sale

Upload your text at www.GRIN.com
and publish for free

ALFRED COSMAS BUTELE

The Host, Habitat and Geographical Range; and Disease Relationships of Venomous and Parasitic Arthropods, and Arthropod-Borne Parasites

A Reference for Tracing Global Warmin and Climate Change Effects

GRIN Verlag

This book at GRIN:

BUTELE COSMAS ALFRED

COURSE TITLE: MEDICAL AND VETERINARY ENTOMOLOGY

The Host, Habitat and Geographical Range; and Disease Relationships of Venomous and Parasitic Arthropods, and Arthropod-Borne Parasites

A Reference for Tracing Global Warming and Climate Change Effects

ATLANTIC INTERNATIONAL UNIVERSITY

HONOLULU, HAWAII

© 28/01/2013

TABLE OF CONTENTS

LIST OF ACRONYMS

AU-IBAR	African Union Interafrican Bureau of Animal Resources
CDC	American Centers for Disease Control and Prevention
e.g.	for example
ECF	East Coast Fever
ed	editor
eds	editors
et al	and other people
FAO	Food and Agriculture Organization of the United Nations
ICIPE	International Center for Insect Physiology and Ecology
Ltd.	Limited
M.Sc.	Master of Science
MCF	Mediterranean Coast Fever
No.	Number
p	page
PhD	Doctor of Philosophy
pp	pages
sp.	one particular species
spp.	several species
USA	United States of America
Vol.	Volume
WHO	World Health Organization
WWH	What-When-How.com

Introduction

Venomous arthropods are those that release a poisonous substance (venom) when disturbed. They release the venom in their defense against intruders. Examples of venomous arthropods are scorpions, wasps, some caterpillars and bees. The word "parasite" is derived from two Greek words, "para", meaning "beside", and "sitos" meaning "food". Therefore, a parasite literally means an organism that is beside another organism for purposes of obtaining food. Adam, *et al.* (1979) defined a parasite as an organism which depends for part of its life or for its entire life on another organism, called the host, from which it obtains food and shelter. According to Smyth (1996), hosts are normally of a different species from their parasites. Parasitism is a kind of adaptation for survival, and in any case, a true parasite should not kill its host, lest it will kill its source of survival and/or kill itself too. Although many parasitic organisms are harmless to the host, others are pathogenic; they cause disease in their hosts, leading to morbidity and death of the host. The parasitic mode of life must have been a survival mechanism developed by certain organisms since the beginning of life on earth, about 4 billion years ago. It must have been impossible for these organisms to survive on their own. It is known that environment can change; presenting different conditions each time and organisms struggle to cope and survive with the change. We are aware that the earth has gone through periods of ice, fire, meteorite strikes, volcanic eruptions, global dust veils, acid rain, and continental upheavals. When such changes occur in the environment, organisms naturally take refuge in safer habitats, hosts or geographical locations. Better still, others change behaviour or transit through an evolutionary process into a new organism in order to fit within the new conditions. So, changes in the environment present with both challenges and opportunities. Organisms which were once free living can become parasitic and those which were once non-pathogenic can become pathogenic. Those which do not have survival mechanisms to cope with the change will disappear from the scene. For example, 99% of the life forms that ever appeared on earth met with extinction already (WWH, 2012). The remaining 1% today consists of an estimated 5-30 million species of flora and fauna. Of these, approximately 1.7 million have been named, and more than half are insects. An insect is that animal as defined and described by Davies (1988), Anonymous (2012a) and Wikipedia (2012a). It is estimated that insects make up 75% (1.3 million) of the known animal kingdom. Insects evolved all through from the time they appeared about 400 million years ago (Devonian period) to occupy every possible land habitat on earth, including freshwaters, except in the depths of saline waters of the sea. Because insects occupy almost every conceivable terrestrial niche, they interact with humans and other organisms, including micro-organisms and domestic animals in countless ways, with varying degree of impact. One case of association of insects with other organisms is parasitism. Such information about insects is obtained through the discipline of Entomology, a study done by scientists called entomologists. The definition of Entomology has been reviewed in Cambridge International Dictionary (1995) and Wikipedia (2012b). Entomological studies have revealed that insects do not exist as stand alone organisms; they co-exist with other closely related organisms. In biological classification and taxonomy, insects (class Insecta) belong to yet a larger group (phylum) of animals called Arthropoda. Arthropoda means 'jointed feet' and includes all animals with segmented legs, segmented bodies and exo-skeleton. Therefore, arthropods include millipedes, centipedes, horseshoe crabs, crayfish, crabs, shrimps, lobsters, sowbugs, pillbugs, scorpions and relatives, spiders, mites, ticks, insects, and many other related organisms. Although entomology is synonymous with the scientific definition of a true insect, it evolved primarily to concentrate on both insects and arachnids. However, some other arthropod classes like Diplopoda, Chilopoda are still being considered by entomologists. Even a few non-arthropod groups like the phylum Mollusca (snails and slugs) and phylum Nematoda (nematodes) are sometimes referred to entomologists (Anonymous, 2012a). Therefore, entomology still encompasses arthropods other than insects as well. Owing to their generally small size, numerical superiority, ability to inhabit every ecological niche on land and freshwaters, and behaviour and activities; arthropods, as a whole, have over the time developed varying associations with organisms, including parasitism, as observed with the insect subgroup earlier. Entomological studies have also revealed that arthropods such as ticks and tumbu fly are parasites in their own right. Others like mosquitoes, tsetse flies carry parasites that cause disease in humans and animals. Some of them like ticks and lice are both parasites and vectors of other parasites and pathogens. Further still, others like ticks are all through being venomous, parasitic and vectors of disease causing parasites. These discoveries all started when Josiah Nott, a mobile, Alabama, physician, proposed (1848) that the causative agents of malaria and yellow fever were transmitted by mosquitoes. In 1881, Carlos Finlay, a Cuban physician, postulated that mosquitoes transmitted the yellow fever agent, setting the stage for Major Walter Reed and associates to verify his claim. However, the foundation for modern medical and veterinary entomology was said to be laid by Louis Pasteur, a French microbiologist who formulated the theory of microbial causation of disease (germ theory), based on his work with the silkworm, *Bombyx mori* in 1887. This was followed by Theobold Smith who in 1889 discovered the causative agent of Texas cattle fever and, working with F. I. Kilbowen, showed in 1893 that the cattle tick, *Boophilus annulatus*, was the vector. In 1897, Ronald Ross demonstrated the occurrence of the malaria parasite in mosquitoes that fed on a human patient, whose blood contained the parasite, thus leading to the elucidation of the epidemiology of malaria. Their work paved the way for the present day research, knowledge and practices of arthropod vector borne diseases and management. Many arthropod vector borne diseases have now been known to occur in different parts of the world, including the human African trypanosomiasis, sometimes called Sleeping Sickness, reported in 37 sub-Saharan African countries (Biryomumaisho, 2007a; and *Science* Daily, 2012). It is caused by a protozoan of genus *Trypanosoma*, transmitted by tsetse flies (*Glossina* spp.), mainly while taking a blood meal from a human host victim. Tsetse flies also transmit a form of trypanosomiasis to domestic animals called animal trypanosomiasis or nagana. Malaria has been confirmed to be a disease transmitted by female Anopheles mosquito to humans, also while taking a blood meal. It is caused by a protozoan of genus *Plasmodium*. Malaria is a killer disease, but difficult to control, yet it continues to be a major public health problem in most countries of the tropical world, with majority of the deaths being reported from Africa (WHO, 1995). Another arthropod vector borne disease is East Coast Fever (ECF), caused by a protozoan organism, *Theileria parva*. It is transmitted by ticks, *Rhipicephalus appendiculatus*, to cattle, mainly in eastern, central and southern Africa. The disease presents with a number of early clinical signs, including reduced milk yield (Anonymous, 2008) with the consequent loss of income to the farmer. Complicated form of ECF is even terminal. Although the emphasis here appears to be on the actual disease parasites they transmit, the biting and blood sucking behaviour of some of the arthropods in itself is being parasitic, as mentioned earlier, should not be taken for granted. From the explanations above, what comes to the mind is that certain arthropods can be very disturbing to human and livestock health and well being. For a long time their management has been based on the theory that they have certain specific hosts, habitats and geographical distribution patterns; knowledge obtained through research. However, with the realities of global warming and climate change on ground, it is widely feared that the parasites will expand their hosts, habitats and geographical locations; thus shifting and enlarging the incidence and distribution of diseases. There will be emergence of new diseases and re-emergence of diseases previously suppressed or eradicated. This calls for taking an account of currently known arthropod-borne parasites and their host, habitat and geographical range. This will help to trace global warming and climate change effects on the parasites and to better understand and manage the dynamics involved in case they take on new hosts, habitats and geographical locations. This paper therefore reviews the host, habitat and geographical range; and disease relationships of venomous and parasitic arthropods, and arthropod-borne parasites.

Description

Many people have experienced the discomfort and pain caused by venomous arthropods. Examples of venomous arthropods are ants, wasps, spiders, fleas, ticks, scorpions. Their venom can cause more than mere pain. Harris (2013) reported that more human deaths in the United States are attributed to venomous arthropods than any other group of venomous animals, including snakes. Although deaths caused by arthropod bites or stings represent only a tiny fraction of the millions of victims, a higher number of about 25,000 per year in the United States have severe reactions. Arthropods often live in close proximity to people and can be abundant, resulting in a high contact rate with people. Arthropods cause three types of envenomization (exposure to venom): piercing/biting, vesicating/urticating, and stinging. Piercing/biting arthropods inject a toxin through their mouth parts, and they include the chiggers, fleas, ticks, spiders, centipedes, wheel bug; millipedes are not venomous, but some of the large tropical species are said to release poisonous or pungent fluids from cutaneous glands when they are irritated. Vesicating/urticating arthropods release toxins on contact through venomous hairs (urticating) or small body openings (vesicating), and they are represented by blister beetles, certain caterpillars with stinging hairs or spines; e.g. Io moth caterpillar, puss caterpillar, and saddleback caterpillar. Stinging arthropods inject a toxin through a stinger located on the posterior end of the abdomen and the group includes paper wasps, hornets, yellowjackets, cicada killers, fire ants, and scorpions. The biology, ecology and systematics of these arthropods have been reviewed by Davies (1988), and Harris (2013). Their host, habitat and geographical range is highlighted in Table 1, but it is important to note that most of them do not have specific host organisms. Fortunately, they are conspicuous organisms, and in a likely event of change of their ecological habitat to escape harsh conditions brought about by global warming and climate change, they could still be noticed in their new ecological habitats, and be avoided or controlled and their venom treated as in Harris (2013). Therefore, more emphasis will be put on the parasitic arthropods and arthropod-borne parasites which hang on other organisms as their hosts for survival. It is important to map out their host, habitat and geographical range. Should the currently known parasitic arthropods and arthropod-borne parasites take on new hosts, habitats and geographical locations, the situation could be more complex and difficult to manage! So, a deeper understanding of them will help to better envisage what would happen if they expand their hosts, habitats and geographical locations. Currently, the widely known malaria causing parasites, plasmodia, belong to phylum Apicomplexa, class Sporozoea, subclass Coccidia, suborder Haemosporina, family Plasmodiidae, and genus *Plasmodium* (Smyth, 1996); while the vector Anopheles mosquito belongs to class Insecta, order Diptera, family Culicidae, and genus *Anopheles* (Davies, 1988). There are 4 main malarial parasites of man: *Plasmodium vivax*, *P. malariae*, *P. falciparum*, and *P. ovale*; but the most virulent being *P. falciparum* (Smyth, 1996). As said earlier, malaria is one of the greatest killer diseases, ranking with cancer and heart disease. Despite continuous research and control programmes in many countries, the situation has shown little improvement, and malaria continues to be a major public health problem in most countries of the tropical world. WHO (1995) reported that of the total world population of about 5.4 people by then, 2200 million were exposed to malaria infection in some 90 countries or areas. It is estimated that there may be 300-500 million clinical cases each year, with countries in tropical Africa accounting for more than 90% of these. Malaria is also said to be the cause of an estimated 1.4-2.6 million deaths worldwide every year, with more than 90% in Africa alone. It is one of the most important causes of mortality and morbidity among infants and young children, and infection during pregnancy contributes, primarily in primiparae, to maternal mortality, as well as to neonatal mortality and low birth weight. The burden put by this disease on household income and human labour force in the affected communities is enormous. Global warming is further compounding the situation. For example, in certain areas like in Kabale District in Southwestern Uganda which used to have cold weather, unfavorable for mosquitoes, are now warming up which has increased incidents of malaria more than previously reported (Environmental Alert, 2010). The need to review and understand the host, habitat and geographical range of the parasite is therefore paramount. Another disease being transmitted by mosquitoes is filariasis, caused by filariae. Although there is no universally agreed classification of nematodes yet, filariae are, for convenience, put in phylum Nematoda, subclass Secernentea (Phasmidea), order Spiururida, suborder Spirurina, and superfamily Filarioidea. The biology of filariae is quite substantially covered by Smyth (1996). A number of the species of filariae are known to parasitize man. In many of the human cases, infections give rise to revolting fleshy deformities which are collectively called elephantiasis. This condition results from inflammation of the walls of the lymphatics and the consequent hyperplasia and partly from mechanical blockage by the worms. Many species of filaria exhibit periodicity. Of the common species of filariae of man, 3 of them responsible for most of the cases of human filariasis are *Wuchereria bancrofti*, *Brugia malayi* and *Onchocerca volvulus*. *Wuchereria bancrofti*, vectored by members of *Culex* (*C.*) *pipiens* "complex" in urban areas and species of *Anopheles*, *Aedes* and more rarely *Mansonia.*; causes bancroftian filariasis, resulting in elephantiasis in man. *Brugia malayi*, resembling *W. bancrofti*, causes Malayan filariasis in man. Approximately 750 million people are at risk of lymphatic filariasis, mainly caused by *W. bancrofti*. Nearly 80 million people are infected and some

30 million of them experience the chronic disease. Of those with chronic infection, more than 1 million suffer from overt elephantiasis, the most disfiguring form of the disease (WHO, 1995). Documentation of the host, habitat and geographical range of these filarial parasites will help in designing better control strategies for filariasis. Dengue and dengue haemorrhagic fever are the most important arboviral diseases, transmitted again by mosquitoes. The main vector in urban areas is *Aedes aegypti* and in suburban and rural areas is *Ae. albopictus*. Vector *Ae. albopictus* is spreading in the world and giving cause for concern. Epidemics of dengue and dengue haemorrhagic fever threaten nearly two-fifths of the world's population, in 100 countries, accounting for millions of cases of disease and thousands of deaths each year. Dengue has recently caused extensive epidemics in non-immune populations in Africa, the Americas, Asia, the Pacific islands and certain countries of WHO's Eastern Mediterranean Region. 37 countries have experienced outbreaks of dengue and dengue haemorrhagic fever, and in many countries, outbreaks of dengue and dengue haemorrhagic fever are the leading cause of hospitalization of young children (WHO, 1995). Highlighting the knowledge of the vector range for the pathogen is crucial. Japanese encephalitis is another viral disease transmitted by mosquito, *Culex tritaenirhynchus*. It has been reported in some countries in Asia and the Pacific islands. Yellow fever is yet another viral disease transmitted by mosquitoes. The number of cases of yellow fever reported to WHO from Africa in the mid-1980s was 5104 but decreased to 2561 cases in 1991. Understanding the vector range of Japanese encephalitis, yellow fever and other mosquito viruses is very important for developing better control strategies in future. Current control options for malaria and other mosquito-borne diseases have been reviewed by WHO (1995) and CDC (2012a). Ticks are one of the parasitic arthropods in their own right while at the same time transmitting disease causing parasites to humans and animals. Description of parasitic arthropods will be detailed in the following section. Tick transmitted diseases are collectively called Tick-borne Diseases (TBDs). Ticks belong to phylum Arthropoda, class Arachnida, order Acarina, suborder Ixodoidea, and 3 families of Argasidae, Ixodidae, and Nuttallielidae. Ticks obtain blood meal from vertebrate or invertebrate host. This has led to transmission of various disease agents from infected animals and humans to uninfected animals and humans. Nuttallielidae has only one species whose economic importance is yet to be known. There are about 170 species of soft ticks (Argasidae) and 700 species of hard ticks (Ixodidae), majority of which are found in Africa. Several of these are known to transmit a wide range of micro-organisms; viruses, rickettsia, bacteria, spirochetes, and also protozoa and nematoda, and are reported to surpass all other arthropods in the number and variety of diseases they transmit to animals and man (Okello-Onen, et al., 1999). The biology, ecology and taxonomy of ticks have been reviewed by FAO (1984a), Okello-Onen, et al. (1999) and Anonymous (2008). Ticks of economic importance to livestock in Africa belong to the family Ixodidae. Only 9 species (from 4 genera) out of a total of over 650 species (from 13 genera) are known to be vectors of economically important diseases. These are *Rhipicephalus* (*R. appendiculatus*), *Boophilus* (*B. decolaratus*, *B. microplus, and B. annulatus*), *Amblyomma* (*A. variegatum, A. hebraeum*) and *Hyalomma* (*H. anatolicum, H. detritum*, and *H. dromedarii*). They transmit a variety of TBDs that can be categorized into protozoan diseases (Theileriosis and Babesiosis), rickettsial diseases (Anaplasmosis and Cowdriosis) and tick-associated dermatophilosis, etc. Although management of

TBDs has been reviewed in FAO (1984b) and Anonymous (2008), there is still need to review the host, habitat and geographical range of ticks and tick-borne parasites. Flagellates of the genus *Leishmania* are parasites of man and other mammals, including dogs. *Leishmania* spp. belong to the phylum Sarcomastigophora, subphylum Mastigophora (Flagellata), class Zoomastigophorea, order Kinetoplastida, suborder Trypanosomatina, family Trypanosomatidae, and genus *Leishmania*. They cause diseases collectively known as leishmaniases, which can be grouped into three: cutaneous leishmaniasis, mucocutaneous leishmaniasis (*espundia*), and visceral leishmaniasis (Kala-azar or black disease or "dum-dum fever" or "ponos"). These are serious debilitating and disfiguring diseases which occur in African, Asian, American, and Mediterranean region countries. All forms of leishmaniasis of man are transmitted by the bite of female sandflies of the subfamily Phlebotominae, which contains about 600 species and subspecies; some 70 of these are proven or suspected vectors of leishmaniasis (Smyth, 1996). Tsetse flies belong to class Insecta, order Diptera, and family Glossinidae. They inhabit sub-Saharan Africa. The biology, ecology, systematics, and distribution of tsetse have been extensively studied and documented by Smart *et al.* (1943), Jordan (1961), Ford (1968), Kangwagye (1968), Kangwagye (1988a), Nash (1969), Locke and Smith (1980), Service (1980), Turner (1980), FAO (1982a), FAO (1982b), Young (1982), Turner (1987), Dransfield (1988), Yu *et al.* (1996), Kalyebi (1998), Vreysen and Khamis (1999), and Mugasa (2007a). 31 species and subspecies of tsetse flies are known to exist in Africa. As already stated in the introduction section, tsetse fly is known to transmit trypanosomes parasitic to both man and livestock, human African trypanosomiasis (sleeping sickness) and animal trypanosomiasis (nagana) respectively. Smyth (1996) grouped trypanosomes in the phylum Sarcomastigophora, subphylum Mastigophora (Flagellata), class Zoomastigophorea, order Kinetoplastida, suborder Trypanosomatina, family Trypanosomatidae, and genus *Trypanosoma*. Between 50,000 and 70,000 people are infected with human African trypanosomiasis (sleeping sickness) and about 60 million are at risk of human African trypanosomiasis (sleeping sickness) infection, as reported in *Science* Daily (2012). Although trypanosomiasis is known to occur in 37 sub- Saharan African countries, it is important to get a renewed picture of the host, habitat and geographical range of the various *Trypanosoma* spp. to guide planners and implementers of human and livestock health projects and programmes. Past and present control options for tsetse and tsetse-borne diseases have been reviewed by Jordan (1986), Kangwagye (1988b), FAO (1982c), FAO (1992), FAO (1993), Biryomumaisho (2007b), Biryomumaisho (2007c), Mugasa (2007b), Waiswa (2007a), Waiswa (2007b), Waiswa (2007c), Waiswa (2007d) and CDC (2012). Chagas' disease is caused by *Trypanosoma cruzi*, transmitted by brightly coloured triatomid bugs belonging to the family Reduviidae, and subfamily Triatominae, all stages of which (larva, nymph and imago) are susceptible to infection (Smyth, 1996). *Trypanosoma cruzi* is passed to the victim through defecation by the vector after a blood meal. Chagas' disease occurs throughout South and Central America; hence it is sometimes referred to as South American Trypanosomiasis. It is essentially a zoonosis, affecting both humans and other wild mammals. The main vectors are known. However, the focus has been on the human pathology, and by 1985 over 24 million people were infected, or at least serologically positive for *T. cruzi* (Smyth, 1996). Do we precisely know the host, habitat and geographical range for *T. cruzi*? *Onchocerca volvulus* causes a disease called onchocerciasis in man. Onchocerciasis is one of the world's distressing diseases of helminth origin, often resulting in blindness, generally referred to as "river blindness" because its main vector, black flies of genus *Simulium* occupy riverine habitats. The population at risk of onchocerciasis infection in the world is 85,583,780 while that in Africa, including Sudan is 80,750,000. The population infected with onchocerciasis in the world is 17,757,700 while that in Africa, including Sudan is 17,640,500. Current options for onchocerciasis and vector control have been reviewed by WHO (1987). Other species of *Onchocerca* have been detected in domestic animals. Fleas belong to class Insecta, order Siphonaptera; and a number of families, subfamilies, genera and 2,400 species are so far known. One important example is the "tropical rat flea", *Xenopsylla cheopis* which transmits bubonic plaque bacillus to man. Dracunculoid worms belong to phylum Nematoda, subclass Secernentea (Phasmidea), order Spiururida, suborder Camallanina. They are parasitic in vertebrates, and vectored by a copepod host. One key example, *Dracunculus medinensis*, sometimes called guinea worm, is infective to man. There are over 140 million people at risk of *Dracunculus medinensis* infection. The annual incidence in Africa is estimated at 3.32 million (Smyth, 1996). It is has been known to be vectored by *Cyclops* sp. Human infection is brought about by accidentally taking infected copepods in drinking water. The disease caused by this parasite in humans is called dracontiasis or dracunculiasis, and some control options have been reviewed by Smyth (1996). *Mansonella perstans*, *M. streptocera*, and *M. azzardi* vectored by midges, *Culicoides* spp. infect man, among other mammals. They cause diseases collectively called mansonellosis in different parts of the world. There are many other parasites (Tables 3) pathogens (Table IV) vectored by arthropods. In general, arthropods of major medical and veterinary importance have been identified among mosquitoes (Culicidae), tsetse flies (Glossinidae), horse flies and deer flies, moth flies, black flies, muscid flies, skin bots, grubs, louse flies, lice, mites, ticks, kissing bugs, bedbugs, Lepidoptera and Hymenoptera, fleas, spiders and scorpions.

General analysis

Major arthropod groups that have so far been found to be parasitic in their own right include lice, *Pediculus humanus* and *Phithrus pubis* (Anoplura: Pediculidae); jigger flea, *Tunga penetrans* (Siphonaptera); mites, *Sarcoptes scabei*; chiggers (Trombiculidae); bot fly (berne), *Dermatobia hominis* (Diptera), tumbu fly, *Cordylobia anthropophaga* (Diptera), and ticks (Acarina) as in Hamilton (2013) and Ghaffar and Hunt (2013) . The kind of parasitism exhibited by the bot fly and the tumbu fly is different from the rest in the list. Their larvae infect and bury themselves in tissues of man, domestic and wild animals; a condition called myiasis. It is an obligatory step in the life cycle of some flies and incidental for others, a phenomenon typical of dipterous larvae. Other species exhibiting myiasis are found in genera *Cochliomyia* (Screw worm fly), *Calliphora*, *Oestrus*, *Sarcophaga,* and *Gastrophilus*. Myiasis may be cutaneous, arterial, intestinal or urinary, in normal tissue or in pre-existing wounds, some of which may result from other infections. Larvae can burrow through necrotic or healthy tissue using their mandibular hooks aided by proteolytic enzymes. They can cause mechanical damage and the affected area may be the site of a secondary infection. Control and treatment for pediculosis (infection by *Pediculus humanus*), phithriasis (infection by *Phithrus pubis*), acariasis (infection by mites), myiasis (infection by dipterous larvae), and tick infection have been reviewed by FAO (1982c), and Ghaffar and Hunt (2013). Arthropods outlined in Table 3 are vectors of parasites. The parasites have been categorized into protozoans, filarial nematodes, trematodes and cestodes. Of the 85 arthropod-borne parasites reported, more than three quarters (66, 78%) are protozoans, 12 filarial nematodes (14%), 2 trematodes (2%), and 5 cestodes (6%). Again of the 85 parasites reported, 70 (82%) of them have been reported to be pathogenic, causing disease; 55 (65%) protozoans, 8 (9%) filarial nematodes, 2 (2%) trematodes, and 5 (6%) cestodes; leaving only 18% as a group of non-pathogenic and those whose pathogenicity has not been established. The later group is represented by 11 protozoans (13%), 4 filarial nematodes (5%), no trematodes (0%), and no cestodes (0%). All the trematodes and cestodes reported are pathogenic. 22 protozoans (26%) are zoonotic; causing diseases to both man and animals, including 5 filarial nematodes (6%), 1 trematode (1%), and 5 cestodes (6%). All cestodes have been observed to be zoonotic. From the above analysis, it is apparent that protozoans are responsible for most of the arthropod-borne diseases in man and livestock, as earlier observed by Adam, *et al* (1979). One protozoan parasite, *Dirofilaria immitis,* a flariid parasite is best known to occur in dogs. Its distribution is now worldwide, although it was long considered to be limited to tropics and subtropics. Some 60 species of mosquitoes are recorded as vectors of this species, and since many of these attack man, it is not surprising to find that man is occasionally infected, with some 100 cases so far reported in literature (Smyth, 1996). Diseases caused by arthropod-borne protozoan parasites include Chagas' disease, a zoonosis, caused by a protozoan, *Trypanosoma cruzi,* a form of trypanosomiasis and transmitted by triatomid bugs to humans mainly in the Americas; African trypanosomiasis, a zoonotic disease (sleeping sickness in man and nagana in cattle), caused by other *Trypanosoma* spp. and vectored by tsetse fly (*Glossina* spp.). Many *Trypanosoma* spp. are still non-pathogenic and vectored by other organisms like triatomid bugs, tabanid flies, fleas and ked. Histomoniasis in birds is another protozoan infection caused by *Histomonas meleagridis* and vectored by nematode worm, *Heterakis gallinarum*; leishmaniasis, a zoonosis, caused by *Leishmania* spp. mainly in man and transmitted by Sandflies, *Phlebotomus* spp. and Lutzomyia sp.; malaria, a zoonotic disease, affecting mainly humans, caused by *Plasmodium* spp. and transmitted by female Anopheles mosquitoes. However, the avian form of malaria is transmitted by mosquitoes of culicinae subfamily. Avian malaria can also be caused by other *Plasmodium*-related protozoans, *Haemoproteus columbae, Leucocytozoon simondi,* and *Leucocytozoon (Akiba) caulleryi* transmitted by hippoboscid flies, black flies, *Simulium* spp. and midges, *Culicoides* spp. respectively. The malaria caused by *Leucocytozoon simondi* in turkey is sometimes called Leucocytozoonosis. Babesiosis, a zoonotic disease, affecting mainly domestic animals and rarely man, is also a protozoan disease caused by various species of *Babesia* and transmitted by various tick species (Acarina). The vectors of some *Babesia* spp. found in pigs and cats are still unknown. *Theileria* spp. cause diseases collectively called theilerioses, including East Coast Fever (ECF), Mediterranean Coast Fever (MCF), Malignant Theileriosis (Ovine Theileriosis), Tropical Theileriosis, and Corridor Diseases, in domestic animals, all also transmitted by various tick species. Protozoans belonging to genus *Anaplasma* cause disease called anaplasmosis in domestic animals; the parasites are transmitted by various tick species. Some *Anaplasma*-related protozoans of genus *Eperythrozoon* and *Haemobartonella* are relatively non-pathogenic except some low degree of pathogenicity exhibited by *Eperythrozoon suis* in pigs. Filarial nematodes, ranking second only to protozoans, are among the major arthropod-borne parasites responsible for a number of diseases in domestic animals and man; including onchocerciasis (river blindness) caused by *Onchocerca volvulus* in man and transmitted by black flies of *Simulium* species. Other arthropod vector borne filarial diseases are Bancroftian filariasis (elephantiasis) in man caused by *Wuchereria bancrofti* and vectored by mosquitoes; dirofilariasis in dogs caused by filarial nematode, *Dirofilaria immitis* and transmitted also by mosquitoes; Loaiasis (eye worm disease) in man caused by *Loa loa* and transmitted by tabanid flies, *Chrysops silacea* and *Chrysops diminiata;* and dracunculosis in man caused by a filarial nematode or guinea worm, *Dracunculus medinensis* and vectored by copepod, *Cyclops* sp. Cestodes and trematodes are also important parasites of man and domestic animals. Cestode (tapeworm), *Diphyllobothrium latum* vectored by fresh water copepods, *Cyclops* sp. is known to cause diphyllobothriasis in man, pigs and dogs. Cestode (tape worm), *Diphylidium caninum* are parasites in dogs, cats, foxes, and man, causing human dipylidiasis mainly in children. The disease is vectored by fleas; dog flea, *Ctenocephalides canis*; cat flea, *C. felis*; human flea, *Pulex irritans*; and dog louse, *Trichodectes canis*. Cestodes (tape worms), *Diphyllobothrium* and *Spirometra* species both cause sparganosis (sparganum infection) in man, cats, dogs, vectored by fresh water copepods, *Cyclops* sp.; cestode (tape worm), *Hymenolepsis diminuta* causes rat tape worm infection in rats, other rodents, other mammals, including man, vectored by fleas, *Xenopsylla cheopis* and other various rodent fleas, *Ceratophyllus fasciatus*, *Ctenocephalides canis*, etc, and beetles, *Tenebrio molitor* (larva), *Tribolium confusum* (larva). Cestode (tape worm or dwarf tape worm), *Hymenolepsis nana,* vectored by flour beetle, is known to cause hymenolepiasis disease in rodents and man. *Paragonimus westermanni*, a trematode, causes paragonimiasis in man, cats and dogs. It is vectored by crustacean crabs and crayfish. Another group of organisms of great importance is the genus *Schistosoma.* Species of *Schistosoma* belong to phylum Platyhelminths, class Trematoda, class Digenea, and family Schistosomatidae. The biology of helminthes in general and schistosomes in particular is reviewed by Smyth (1996). Five species of *Schistosoma* are known to be pathogenic parasites of man. Of these, the chief ones are *Schistosoma mansoni, S. haematobium,* and *S. japonicum,* with *S. mekongi* and *S. intercalatum* having limited distribution. Other species, *S. matheei* and *S. bovis*, are occasionally parasites of man and *S. incognitum* may also prove to be infective to humans. The disease caused by schistosomes called schistosomiasis (bilharzias) is the most important disease of helminth origin and causes untold misery in some 75 countries. Some 200 million people are probably infected and 500-600 million more are exposed to infection (Smyth, 1996). However, schistosomiasis is not transmitted by an arthropod vector, but rather by a molluscan host, snails, of genus *Biomphalaria.* Human infection with schistosomiasis is brought about by bathing or wading in infected waters. Control of schistosomiasis depends basically on the control of the snail vectors, a problem which presents an extremely complex ecological situation. A summary of the above diseases appear in Table 3. Pathogens other than parasites that cause diseases in animals and man and are also vectored by arthropods have been listed separately in Table IV. They have been categorized into bacterial/rickettsial and viral pathogens. The diseases they cause include, among others, Rocky Mountain Spotted Fever, Lime disease, tularemia, human anthrax, Scrub Typhus (Tsutsugamushi disease) , Colorado Tick Fever (CTF), yellow fever, Nairobi sheep disease, St. Louis Encephalitis, and dengue fever.

Actualization

The Host, Habitat and Geographical Range, and Disease Relationships of Venomous Arthropods

Venomous arthropods include ticks, chiggers, cat fleas, spiders, centipedes, wheel bugs, blister beetles, Io moth caterpillars, puss caterpillars, saddleback caterpillars, paper wasps, hornets, yellowjackets, cicada killers, fire ants, and scorpions (see Table 1 below). The effects of their venom on the victim's body range from pains through skin irritations, scratching & scarring (wounds), inflammatory body reactions (swelling), allergy, asthma, toxicosis to death. Information on venomous arthropods by country has been scanty; much of the geographical distribution has been generalized at continental level. However, for immediate use for precautionary measures, the information about their ecological habitats provided here will suffice. Importantly, most of them do not have specific hosts, except ticks, chiggers and cat fleas. Ticks are widespread; particularly in Africa, Asia, the Mediterranean Region, and America. They either actively search or wait to attach to (ambush) their passing host mainly in overgrown grass, weeds and wooded or shrubby vegetation (Harris, 2013). They have a wide range of hosts, including mainly cattle, horses, sheep, goats, dogs, pigs and man. Other hosts are donkeys, mules, camel, and wild animals such as antelopes, zebra, buffalo, and warthogs. In the animal host, adults can be found on the sides of the body, shoulders, neck, dewlap, sometimes the whole body; immature stages on the tips of upper edges of the ears, legs, belly or dewlap; on human host, they get attached to the base of the scalp, waist, knee or armpit. Chiggers have reported from South Asia, USA, and Africa. The larvae crawl up grass blades, weeds or other vegetation, especially tall grass at the edge of lawns, or in densely vegetated areas bordering lawns, playgrounds, parks or golf roughs and attach themselves to a passing host (human or animal). Cat fleas are said to be common in USA and Europe. These fleas live in pet resting places, such as areas close to building foundations, under decks and porches, doghouses and kennels. The host range includes cats, dogs and human; wild hosts, stray cats and dogs, raccoons, opossums, moles. You can find fleas all over the body of the infected host. The rest of the venomous arthropods do not have specific hosts as said earlier. Their ecological habitats have been highlighted in Table 1 to guide ground workers and other professionals whose work involve interfering in their habitats so that they can avoid their attack and venom.

Table 1: A Summary of the Host, Habitat and Geographical Range, and Disease Relationships of Venomous Arthropods

S/No.	Species	Host range	Habitat (Predilection site) Range	Geographical Range	Disease relationship as a venomous arthropod, minus parasites and pathogens
1.	Ticks	Passing host: cattle, horses, sheep, goats, dogs, pigs, man, donkeys, mules, camel, and wild animals such as antelopes, zebra, buffalo, and warthogs.	Found in tall grass, weeds and wooded or shrubby vegetation in moist and humid areas; in the animal host, adults can be found on the sides of the body, shoulders, neck, dewlap, sometimes the whole body; immature stages on the tips of upper edges of the ears, legs, belly or dewlap; on human host, they get attached to the base of the scalp, waist, knee or armpit	Widespread: particularly in Africa, Asia, the Mediterranean Region, and America.	Skin irritation at the site of bite
2.	Chiggers (larvae of mites sometimes referred to as "red bugs" or "harvest mites.")	Passing human or animal	Found on grass blades, weeds or other vegetation; especially tall grass at the edge of lawns, or in densely vegetated areas bordering lawns, playgrounds, parks or golf roughs, host's whole body	South Asia, USA, Africa	Itching around the infected area.
3.	Cat fleas	Cats, dogs and human; wild hosts, stray cats and dogs, raccoons, opossums, moles, etc.	Found in pet resting places, such as areas close to building foundations, under decks and porches, doghouses and kennels; host's whole body	Europe, USA	Skin irritation
4.	Spiders	Not specific; accidental contact with	Found in outdoor areas: in the web under stones, woodpiles, loose	World-wide: Africa, Asia, America	Pain in the affected area

			bark, storage buildings, barns, outhouses and water faucets.		
5.	Centipedes	Not specific; accidental contact with man or children unknowingly handling them	During the day: can be found under rocks, boards or bark and in crevices, basements, porches, patios and moist, protected areas. They become active at night to feed on prey such as small insects.	Widespread; including Africa, Asia, South America, and Mediterranean areas	Venom and burning pain at the affected area, body reaction
6.	Wheel bug	Not specific; accidental contact with man	Found in vegetation, boards, rocks or other objects	Widespread	Toxic painful bite
7.	Blister beetles	Not specific; accidental encounter with man	Not specific; mobile	Widespread	Cantharidin and blister formed on the skin
8.	Io moth caterpillar	Not specific; accidental encounter with man	Can be found feeding in shrub	Widespread, including Africa and America	Toxic spines and burning pain in the affected area, inflammatory body reaction
9.	Puss caterpillar ("asps")	Not specific; accidental encounter with man	Can be found feeding in shrub	Widespread, including Africa and America	Burning pain in the affected area, inflammatory body reaction
10.	Saddleback caterpillar	Not specific; accidental encounter with man	Can be found feeding in shrub	Widespread, including Africa and America	Toxic spines and burning pain in the affected area, inflammatory body reaction
11.	Paper wasps	Not specific; accidental contact with man	Found in umbrella-shaped nests attached to buildings, trees and shrubs by a single stalk	Widespread, including Africa, America, Europe and Asia	Burning pain at the site of sting, inflammatory body reaction
12.	Hornets	Not specific; accidental contact with man	Found in large multi-combed nests made of gray paper-like material resembling a large football in shrubs, trees, on overhangs of buildings and other structures.	Widespread	Burning pain at the site of sting, inflammatory body reaction
13.	Yellowjackets	Not specific; accidental	Found in nests underground in	Widespread, including	Burning pain at the site of sting,

		contact with man	burrows or under rocks or landscape timbers, attics, sheds or other structures; also found scavenging for food outdoors around picnic areas, parks and similar sites	Africa and America	inflammatory body reaction
14.	Cicada killers	Not specific; accidental contact with man	Burrow into turf and produce a frightening, loud buzz that annoys people; also found searching for cicadas in shrubs and trees; prefer areas with sparse vegetation for nesting	Widespread	Burning pain at the site of sting, inflammatory body reaction
15.	Fire ants	Not specific; accidental contact with humans and animals	Found in unsightly mounds on lawns, playgrounds, golf courses and other landscape areas	Widespread, particularly in South America and Africa	Painful stings, scratching stung areas, secondary infection, scarring, allergic body reactions in humans and animals
16.	Scorpions	Not specific; accidental contact with humans	Found living outdoors under boards, rocks, stones or other structures; active at night	Widespread, including Africa and America	Painful sting which can become persistently sore and swollen

Sources: Okello-Onen, *et al.* (1999); and Harris (2013).

The Host, Habitat and Geographical Range, and Disease Relationships of Parasitic arthropods

Arthropods which are parasitic in their own right include also ticks, of various genera and species. As observed earlier, ticks either actively search or wait to attach to (ambush) their passing host mainly in overgrown grass, weeds and wooded or shrubby vegetation (Harris, 2013). Although certain species specific differences have been observed in precipitation levels in the various habitats, most of the species prefer areas with relatively moderate to high rainfall, moisture content and humidity. Few exceptions to this are only few and specially adapted to live in arid and desert areas; but still, they will attack their hosts from the available vegetation. They have a wide range of hosts, including mainly cattle, horses, sheep, goats, dogs, pigs and man. Other hosts are donkeys, mules, camel, and wild animals such as antelopes, zebra, buffalo, and warthogs. There have also been observed species specific host differences. Their geographical distribution range has been seen to be wide, particularly in Africa, Asia, the Mediterranean Region, and America. The habitat in the host (predilection site) varies with age; adults attach at different sites from the immature stages (larvae and nymphs). These also vary with species. It is advisable to follow the host, habitat and geographical range of each species, as summarized in Table II below, as species may vary from place to place. Minus the venom, parasites and other pathogens (bacterial, rickettsial, viral or fungal) ticks carry, their impact on livestock production and human health will be reduced to only blood loss, tick worry, hide damage, abscesses, cutaneous stretothricosis, tick wounds, unpleasant sores, and production losses in livestock. Cost of treating tick-borne diseases would be zero if not minimal. Tick burden is worsened by adding the venom, parasites and pathogens they vector. Other parasitic arthropods whose host, habitat and geographical range, and disease relationship has been described include, human body louse, *Pediculus humanus humanus*; human head louse, *Pediculus humanus capitis*, human pubic louse, *Phithrus pubis*; jigger flea, *Tunga penetrans*; botfly, *Dermatobia hominis*; tumbu fly, *Cordylobia anthropophaga*; house fly, *Musca domestica*; African face fly, *Musca* spp.; nuisance biting fly (stable fly), *Stomoxys calcitrans*; day biting midge, *Leptoconops* spp.; biting (red) lice, *Damalinia bovis*; and sucking (blue) lice, *Linognathus* spp. (see Table II below).

Table II: A summary of the Host, Habitat and Geographical Range, and Disease Relationships of Parasitic arthropods

S/No	Species	Host range	Habitat Range	Geographical Range	Disease relationship
1.	Body louse, *Pediculus humanus humanus* (Anoplura: Pediculidae)	Man	Body and clothing	World-wide	Pediculosis: Itching of the scalp, leading to dermatitis; fever; aches; and secondary infections in man.
2.	Head louse, *Pediculus humanus capitis* (Anoplura: Pediculidae)	Man	Hair; often back of the neck, behind the ears	World-wide	Pediculosis: Matted hair; itching of the scalp, leading to dermatitis; fever; aches; and secondary infections in man
3.	Pubic louse (crab louse), *Phithrus pubis* (Anoplura: Pediculidae)	Man	Pubic region, armpits, beard, moustache in adults; eyebrows and eyelashes in children; hair. Infection occurs through contact with infected bedding or other objects or sexual contact (veneral)	World-wide	Phthriasis: Matted hair; itching of the scalp, leading to dermatitis; fever; aches; and secondary infections.
4.	Jigger flea (chigger, jigger, chigoe, bicho do pé or sand flea), *Tunga penetrans*	Man, dog; not host specific	Skin under the toenails, fingernails and the sole of the foot in man; and in the paws, under the pads, and the scrotum in dog.	Tropical and subtropical regions of the Americas and Africa	Tungiasis in man

		(Siphonaptera)			
5.	Scabies mite, *Sarcoptes scabei.*	Man	Skin on the webbing side of fingers, later spreading to the wrists, elbows and the rest of the body the buttocks, women's breasts and external genitalia.	World-wide	Acariasis in man
6.	Chiggers, the larvae of red mites or harvest mites, (Trombiculidae)	Man, after walking through a grassy environment.	Attaches to the skin in the ankles, waistline, armpits and perianal area	South Asia	Piercing the skin near a hair follicle; digesting and eating up of skin cells: reaction to the mouth parts and saliva of the mite; erythematous papule appears that is highly pruritic; the intensity of the eruption depends on the sensitivity of the host and may be followed by fever
7.	Bot fly (berne), *Dermatobia hominis* (Diptera)	Mosquitoes, man, domestic and wild animals	Eggs on the abdomen of a mosquito; sub-cutaneous tissue of man, domestic and wild animals.	Mexico and Belize) and south America and some Caribbean islands such as Trinidad,	Myiasis in humans (furunculous disease), domestic and wild animals
8.	Tumbu fly, *Cordylobia anthropophaga* (Diptera)	Man, domestic and wild animals	Eggs on the ground or deposited on clothes; after hatching the larva burrows into the skin	Tropical or sub-Saharan Africa	Furuncular myiasis in humans and animals
9.	Blue tick, *Boophilus* spp.	Cattle, horses, sheep, goats, donkeys; wild animals antelopes, giant eland, deers, Jackson's hartebeest; and man	Tall grass, weeds and wooded or shrubby vegetation in moist and humid areas; in the animal host, adults can be found on the sides of the body, shoulders, neck, dewlap, sometimes the whole body; immature stages on the tips of upper edges of the ears, legs, belly or dewlap; on human host, they get attached to the base of the scalp, waist, knee or armpit	East Africa, Madagascar, Southern Africa, West Africa, Middle East, Mediterranean Region, Southern and Central Sudan	Red water (*Babesia bigemina*); tick-borne gall sickness (bovine anaplasmosis); spirochaetosis of cattle, horses, sheep and goats; *Babesia bovis*, blood loss, tick worry, production losses in livestock; tick bite fever in man
10.	Red-legged tick, *Rhipicephalus evertsi evertsi*	Horses, cattle, sheep, goats, pigs, dogs; wild animals like antelopes, zebra; and man	They are found in vegetated moist areas; on the animal host's body, adults are under tail, around the anus, scrotum, teats larvae and nymphs in ear passages; predilection sites on the human host are same as for the blue tick,	Widely distributed in East Africa, less common in dry areas and coastal regions	*Theileria mutans*, red water of cattle (*Babesia bigemina*), Biliary fever of horses and other equines (*B. equi*), East Coast Fever, spirochaetosis of cattle and horses (*Borrelia theileri*), blood loss, paralysis in lambs, production losses in livestock; tick bite fever in

No.	Tick	Hosts	Habitat / predilection sites	Distribution	Diseases / effects
			Boophilus spp.		man
11.	Brown ear tick, *Rhipicephalus appendiculatus*	Cattle, sheep, goats, horses, mules, donkeys and dogs	They are found in thick vegetation of areas which are moist most of the time of the year; on the animal host's body, both adults, nymphs and larvae attach to the ears, whole head, around eyes, muzzle and ears, base of ears, head poll, tail switch; predilection sites on the human host are same as for the blue tick, *Boophilus* spp.	Areas with adequate vegetation and moist all year round	East Coast Fever, Corridor disease, tick toxicosis, blood loss, and production losses in livestock
12.	Bont-legged tick, *Hyalomma* spp.	Cattle, sheep, goats, horses, camel, dogs; wild animals like buffalo, large antelopes, hares and birds, hedgehogs, burrowing rodents, lions, zebras, rhinos, warthogs, reptiles or smaller animals; and man	Can be found in grassy areas in semi-arid to arid savanna; on the animal host's body, you can find adults on the udder or scrotum, the groin, around the anus, in the tail tuft, between hooves, lower part of dewlap, tail tip, flanks, genitals, perianal regions, heels, ventral part of the body, neck, legs; nymphs on the crown of the head of avian hosts; predilection sites on the human host are same as for the blue tick, *Boophilus* spp.	Whole of Africa	Sweating sickness, hide damage, abscesses, blood loss, *Theileria annulata* in cattle, *T. Camelensis* in camels, equine piroplasmosis (*Nuttallia equi*) in horses, *Babesia equi, B. cabalii, Anaplasma marginale, Trypanosoma theileri,* biliary fever in horses, foot-rot in sheep, production losses in livestock; and tick bite fever, Crimean-Congo haemorrhagic fever, tick paralysis, Q fever, typhus (*Rickettsia conorii*), and plaque (*Pasturella pestis*) in man
13.	Bont tick, *Amblyomma hebraeum*	Cattle, sheep, horses, donkeys, pigs; wild animals reptiles, birds, mammals; and man	Vegetation in lowland savanna with moderate to high rainfall; on the animal host's body, adults on underside of the body, nymphs on feet, belly and groin; larvae on feet and legs; predilection sites on the human host are same as for the blue tick, *Boophilus* spp.	South Africa, Zimbabwe and Mozambique	Heart water, *Theileria mutans, T. velifera, Coxiella burneti* and *Rickettsia conorii*, cutaneous streptothricosis, hide damage, abscesses, cutaneous stretothricosis, blood loss, tick wounds, production losses in livestock; unpleasant sores and Q fever in man
14.	soft tick, *Ornithodoros* spp.	Domestic animals; wild animals, warthog, pigs, porcupine, aardvark, antbears,	Not specific	Wide spread	Toxicosis, anaemia, sensitivity, itching in livestock; swine fever in pigs; African tick-borne relapsing fever, salmonellosis, rickettsiosis, Q fever, typhus, tularemia, leptospirosis in man

		some reptiles; and man			
15.	Housefly, *Musca domestica*	Man and domestic animals	Whole body and face	World-wide	Cholera, typhoid, mastitis, anthrax, tuberculosis, conjunctivitis, production losses
16.	African face fly, *Musca* spp.	Man and domestic animals	Face, wounds	World-wide	Parafilaria, eye worms, pink eye, bacillary dysentery, production losses
17.	Biting fly e.g. Stable fly, *Stomoxys calcitrans*	Man and domestic animals	Body, flanks, legs	World-wide	Tick borne gall sickness (anaplasmosis), bacterial and fungal diseases, intense irritation, production losses
18.	Midges e.g. day biting midges, *Leptoconops* spp.	Man and domestic animals	Ears, along topline	World-wide	Sharp burning sting, intense irritation, production losses
19.	Biting lice (red lice), *Damalinia bovis*	Domestic animals	Shoulders, along topline, tail base and neck	World-wide	Irritation/restlessness, hair loss, unthriftiness, scurfy, production losses
20.	Sucking lice (blue lice), *Linognathus* spp.	Domestic animals	Brisket, chest, shoulders, dewlap and escutcheon	World-wide	Anaemia, irritation/restlessness, hair loss, unthriftiness, scurfy, production losses

Sources: Smith Kline Beecham Chart of some important ecto-parasites; Okello-Onen, *et al.* (1999); and Ghaffar and Hunt (2013).

The Host, Habitat and Geographical Range, and Disease Relationships of Arthropod-Borne Parasites

Arthropod-borne parasites together with the diseases caused have already been analyzed in the preceding section on general analysis. Focus here now is on the host, habitat and geographical range of these parasites. Protozoan parasites of genus *Trypanosoma* have probably the highest number of major host groups, including horses, donkeys, mules, camels, cattle, goats, pigs, dogs, various wild ungulates, sheep, man, buffalo, zebra, warthogs, antelopes, cats, monkeys, opossums, rats, mice, and rabbits. However, in the body of the host they are commonly found in the blood, rarely in muscle tissue especially the cardiac muscle. They cause diseases to both man and livestock called trypanosomiasis, and according to WHO (1995) human trypanosomiasis (sleeping sickness) has been reported from 37 countries in Sub Saharan Africa, namely Angola, Benin, Botswana, Burkina Faso, Burundi, Cameroon, Central African Republic, Chad, Congo, Cote d'Ivoire, Equatorial Guinea, Ethiopia, Gabon, Gambia, Ghana, Guinea, Guinea-Bissau, Kenya, Liberia, Malawi, Mali, Mozambique, Namibia, Niger, Nigeria, Rwanda, Senegal, Sierra Leone, Togo, Uganda, United Republic of Tanzania, Zaire, Zambia, Zimbabwe, Somalia, Sudan, and Syrian Arab Republic. From here any new reports can be traced and be linked to factors that may be biological, technical, social, economic or even political, including global warming and climate change. Probably the second highest ranking parasites in number of major host groups are of filarial origin. The range of hosts include rats, squirrels, man, dog, cats, wolves, foxes, monkeys, pangolins, apes, horses, cattle, antelopes, and chimpanzees. However, filarial worms appear to rank very high in the number of habitat range in the body of their hosts; they can be found in pleural cavity of the host, in the peritoneal space, lung, muscle fibres, cardiac muscle, blood, heart, lymph nodes, lymphatics, subcutaneous tissue, eye, body cavities, tissues, under the skin, between muscles, ligaments, tendons, nodules (onchodermata) under the skin. They cause disease called filariasis (elephantiasis), which has been reported in 70 countries, namely Angola, Benin, Burkina Faso, Burundi, Cameroon, Cape Verde, Central African Republic, Chad, Comoros, Congo, Cote d'Ivoire, Equatorial Guinea, Ethiopia, Gabon, Gambia, Ghana, Guinea, Guinea-Bissau, Kenya, Liberia, Madagascar, Malawi, Mali, Mauritius, Mozambique, Niger, Nigeria, Reunion, Sao Tome & Principe, Senegal, Seychelles, Sierra Leone, Togo, Uganda, United Republic of Tanzania, Zaire, Zambia, Zimbabwe, Brazil, Costa Rica, Dominican Republic, Guyana, Haiti, Suriname, Trinidad & Tobago, Egypt, Bangladesh, India, Indonesia, Myanmar, Nepal, Sri Lanka, Thailand, Brunei Darussalam, Cambodia, China, Cook Islands, Fiji, French Polynesia, Lao People's Democratic Republic, Malaysia, Papua New Guinea, Philippines, Republic of Korea, Samoa, Solomon Islands, Tonga, Vanuatu, and Viet Nam. Another case is that of *Plasmodium* and *Plasmodium*-related malaria causing protozoan parasites. They appear not to rank high in host range, which includes man, chimpanzees, gibbon, domestic fowl, hen, ducks, and geese. Their habitat range in the hosts is not wide either, including blood, placenta, brain, liver, lungs, spleen and tissues; but human malaria has been reported in 93 countries in the world, namely Angola, Benin, Botswana, Burkina Faso, Burundi, Cameroon, Central African Republic, Chad, Comoros, Congo, Cote d'Ivoire, Equatorial Guinea, Ethiopia, Gabon, Gambia, Ghana, Guinea, Guinea-Bissau, Kenya, Liberia, Madagascar, Malawi, Mali, Mauritania, Mozambique, Namibia, Niger, Nigeria, Rwanda, Sao Tome & Principe, Senegal, Sierra Leone, South Africa, Swaziland, Togo, Uganda, United Republic of Tanzania, Zaire, Zambia, Zimbabwe, Argentina, Belize, Bolivia, Brazil, Colombia, Costa Rica, Dominican Republic, Ecuador, El Salvador, French Guiana, Guatemala, Guyana, Haiti, Honduras, Mexico, Nicaragua, Panama, Paraguay, Peru, Suriname, Venezuela, Afghanistan, Djibouti, Egypt, Iran, Iraq, Morocco, Oman, Pakistan, Saudi Arabia, Somalia, Sudan, Syrian Arab Republic, United Arab Emirates, Yemen, Turkey, Bangladesh, Bhutan, India, Indonesia, Myanmar, Nepal, Sri Lanka, Thailand, Cambodia, China, Lao People's Democratic Republic, Malaysia, Papua New Guinea, Philippines, Solomon Islands, Vanuatu, and Viet Nam (WHO, 1995). See Table 3 below for a summary of the host, habitat and geographical range of the major arthropod-borne parasites.

Table 3: A summary of the Host, Habitat and Geographical Range, and Disease Relationships of Arthropod-Borne Parasites

S/No.	Parasite	Host Range	Habitat Range	Vector (s)	Geographical Range	Disease Relationship
(a) Protozoan parasites						
1.	Protozoan, *Trypanosoma* (*Trypanozoon*) *brucei* (Zoomastigophorea: Trypanosomatidae)	Horse, donkey, mule, camel, cattle, sheep, goat, pig, dog and various wild ungulates	Blood	Tsetse fly, *Glossina* spp. (Diptera: Glossinidae)	Tropical Africa (within latitudes 14^0 N and 30^0 S)	Pathogenic: Nagana in domestic animals
2.	Protozoan, *Trypanosoma* (*Trypanozoon*) *rhodesiense* (Zoomastigophorea: Trypanosomatidae)	Man, horse, donkey, mule, camel, cattle, sheep, goat, pig, dog and various wild ungulates	Blood	Tsetse fly, *Glossina* spp. (Diptera: Glossinidae)	East Africa	Pathogenic (& zoonotic): Rhodesian sleeping sickness in man (acute and fatal if not treated)
3.	Protozoan, *Trypanosoma* (*Trypanozoon*) *gambiense* (Zoomastigophorea: Trypanosomatidae)	Man and domestic pig	Blood	Tsetse fly, *Glossina* spp. (Diptera: Glossinidae)	West Africa and Uganda (shores of Lake Victoria)	Pathogenic (& zoonotic): Gambian sleeping sickness in man (chronic, persisting for 1 or 2 years before death)
4.	Protozoan, *Trypanosoma* (*Trypanozoon*) *evansi* (Zoomastigophorea: Trypanosomatidae)	Horse, donkey, mule, camel, cattle, buffalo, sheep, goat, pig, dog and various wild animals	Blood	None; transmitted mechanically by haematophagous insects, the tabanids	Cosmopolitan (tropical and subtropical); particularly in North Africa, southern part of Asia, Indonesia, Philippines and South America.	Pathogenic: Surra in Africa and Asia; and Derrendgadera or Murrina in South America
5.	Protozoan, *Trypanosoma* (*Trypanozoon*) *equinum* (Zoomastigophorea: Trypanosomatidae)	Horse, donkey, mule, camel, cattle, buffalo, sheep, goat, pig, dog and various wild animals	Blood	None; transmitted mechanically by haematophagous insects, the tabanids; and vampire bats	Central and South America	Pathogenic: Mal de Caderas
6.	Protozoan, *Trypanosoma* (*Trypanozoon*) *equiperdum* (Zoomastigophorea: Trypanosomatidae)	Horse, donkey	Blood	None; veneral transmission at copulation	South Europe, Asia, North Africa	Pathogenic: Dourine in Equidae
7.	Protozoan, *Trypanosoma*	Cattle, sheep,	Blood	Tsetse fly, *Glossina* spp.	Tropical Africa	Pathogenic: Nagana in domestic animals

	(Nannomonas) congolense (Zoomastigophorea: Trypanosomatidae)	zebras, warthogs, horse, camels, goats and various wild ungulates		(Diptera: Glossinidae)		
8.	Protozoan, *Trypanosoma (Nannomonas) dimorphon* (Zoomastigophorea: Trypanosomatidae)	Cattle, sheep, equines, pigs	Blood	Tsetse fly, *Glossina* spp. (Diptera: Glossinidae)	Tropical Africa	Pathogenic
9.	Protozoan, *Trypanosoma (Nannomonas) simiae* (Zoomastigophorea: Trypanosomatidae)	Warthogs, pigs, camels, possibly cattle, equines	Blood	Tsetse fly, *Glossina* spp. (Diptera: Glossinidae)	Tropical Africa	Pathogenic
10.	Protozoan, *Trypanosoma (Nannomonas) suis* (Zoomastigophorea: Trypanosomatidae)	Pigs	Blood	Tsetse fly, *Glossina* spp. (Diptera: Glossinidae); and haematophagous insects, the tabanids	Zaire, Tanzania	Pathogenic: fulminating and rapidly fatal disease in domestic pigs
11.	Protozoan, *Trypanosoma (Duttonella) vivax* (Zoomastigophorea: Trypanosomatidae)	Cattle, sheep, equines, goats, antelopes, dogs	Blood	Tsetse fly, *Glossina* spp. (Diptera: Glossinidae)	Tropical Africa	Pathogenic: Souma in cattle (severe)
				Haematophagous insects, the tabanids	West Indian islands, Mauritius and Antilles; and, Central and South America)	
12.	Protozoan, *Trypanosoma (Duttonella) uniforme* (Zoomastigophorea: Trypanosomatidae)	Antelopes, cattle, sheep, goats	Blood	Tsetse fly, *Glossina* spp. (Diptera: Glossinidae)	Central and East Africa, Angola	Pathogenic
13.	Protozoan, *Trypanosoma cruzi* (Zoomastigophorea: Trypanosomatidae)	Man, dogs, cats, armadillos, opossums, raccoons	Blood, muscle tissue (cardiac muscle)	Kissing assassin (Triatomid bugs), *Triatoma* species, *Panstrongylus* spp. (Hemiptera: Reduviidae)	Central and South America, Southern US	Pathogenic (& zoonotic): Chagas' disease in man
14.	Protozoan, *Trypanosoma rangeli*	Man, monkeys, dogs,	Blood	Triatomid bugs	Central and south America,	Non-pathogenic

	(Zoomastigophorea: Trypanosomatidae)	opossums			southern US	
15.	Protozoan, *Trypanosoma theileri* (Zoomastigophorea: Trypanosomatidae)	Cattle, antelopes	Precisely unknown habitat; scanty in blood	Tabanid flies	World-wide	Non-pathogenic
16.	Protozoan, *Trypanosoma lewisi* (Zoomastigophorea: Trypanosomatidae)	Rats	Blood	Flea, *Ceratophyllus fasciatus* and other fleas: the North American and European rat flea, *Nosopsyllus fasciatus*; human flea, *Pulex irritans*; the Indian rat flea, *Xenopsylla cheopis* or the dog flea, *Ctenocephalides canis.*	World-wide	Non-pathogenic
17.	Protozoan, *Trypanosoma melophagium* (Zoomastigophorea: Trypanosomatidae)	Sheep	Blood	Ked, *Melophagus ovinus*	World-wide	Non-pathogenic
18.	Protozoan, *Trypanosoma musculi* (Zoomastigophorea: Trypanosomatidae)	Mice	Blood	Fleas	World-wide	Non-pathogenic
19.	Protozoan, *Trypanosoma nabiasi* (Zoomastigophorea: Trypanosomatidae)	Rabbits	Blood	*Spilopsyllus cuniculi*	World-wide	Non-pathogenic
20.	Protozoan, *Trypanosoma theodori* (Zoomastigophorea: Trypanosomatidae)	Goats	Blood	*Lipoptena capreoli*	Palestine	Non-pathogenic
21.	Protozoan, *Histomonas meleagridis*	Turkey, hen, quail and ruffed grouse	Caecum, liver	Nematode worm, *Heterakis gallinarum*	World-wide	Pathogenic: Histomoniasis (enterohepatitis or ("blackhead") in turkey
22.	Protozoan, *Leishmania donovani* (Zoomastigophorea: Trypanosomatidae)	Man, dog and wild animals, particularly rodents	Blood, lymph gland, bone marrow, spleen, reticulo-endothelial tissue	Sandfly, *Phlebotomus* spp.	All continents, except Australia	Pathogenic (& zoonotic): Visceral leishmaniasis (kala-azar or dum-dum fever or ponos) in man

No.	Parasite (Classification)	Host	Location in host	Vector / Transmission	Distribution	Remarks
23.	Protozoan, *Leishmania tropica* (Zoomastigophorea: Trypanosomatidae)	Man, dog, wild animals, particularly rodents	Blood, lymph gland, skin	Sandfly, *Phlebotomus* sp.; direct human to human transmission also occurs	Old world: Asia, Africa and Europe	Pathogenic (& zoonotic): Cutanec leishmaniasis (bouton d'Orient, oriental sore, Baghdad boil or Delhi sore or Alep button) in man
24.	Protozoan, *Leishmania braziliensis* (Zoomastigophorea: Trypanosomatidae)	Man, dog, wild animals particularly rodents	Blood, lymph gland, nasopharynx, nasal cavity	Sandfly, *Lutzomyia* sp.	New World: North and South America	Pathogenic (& zoonotic): Mucocutaneous leishmaniasis (espundia or American leishmaniasis or chiclero ulcer) in man
25.	Protozoan, *Plasmodium vivax* (Sporozoea: Plasmodiidae)	Man	Blood, liver	Mosquito: *Anopheles* spp.	World-wide	Pathogenic: Benigr tertian malaria in man
26.	Protozoan, *Plasmodium malariae* (Sporozoea: Plasmodiidae)	Man, chimpanzee, gibbon	Blood, liver	Mosquito: *Anopheles* spp.	World-wide	Pathogenic (& zoonotic): Quartan malaria in man
27.	Protozoan, *Plasmodium ovale* (Sporozoea: Plasmodiidae)	Man	Blood	Mosquito: *Anopheles* spp.	West Africa, America, Asia, Philippines	Pathogenic: mild malaria in man
28.	Protozoan, *Plasmodium falciparum* (Sporozoea: Plasmodiidae)	Man, chimpanzee	Blood, liver, placenta, brain, lungs	Mosquito: *Anopheles* spp.	World-wide	Pathogenic (& zoonotic): Malignar tertian malaria in man
29.	Protozoan, *Plasmodium (Haemamoeba) gallinaeceum* (Sporozoea: Plasmodiidae)	Domestic and jungle fowl	Blood, brain	Mosquitoes of culicinae subfamily.	South and South East Asia	Pathogenic: Avian malaria in domestic hens
30.	Protozoan, *Plasmodium (Novyella) juxtanucleare* (Sporozoea: Plasmodiidae)	Domestic hen, turkey, jungle fowl	Blood	Mosquito: *Culex* spp.	Central and South America, Ceylon, Malaya, Japan	Pathogenic: Avian malaria in domestic hen
31.	Protozoan, *Haemoproteus columbae* (Sporozoea: Haemoproteidae)	Domestic and wild pigeons, reptiles	Blood, lung, liver, spleen	Hippoboscid flies	World-wide	Pathogenic: Avian malaria-like condition in pigeor called pigeon mal
32.	Protozoan, *Leucocytozoon simondi*	Ducks and geese	Blood, liver, spleen	Blackflies, *Simulium* spp.	Canada, North America	Pathogenic: Leucocytozoonosi (turkey malaria or

	(Sporozoea: Haemoproteidae)				gnat fever)
33.	Protozoan, *Leucocytozoon (Akiba) caulleryi* (Sporozoea: Haemoproteidae)	Domestic hen	Blood, tissues, including the brain	Midges, *Culicoides* spp.	South East Asia	Pathogenic: Avian malaria in domestic hens
34.	Protozoan, *Babesia bigemina* (Sporozoea: Babesiidae)	Cattle	Blood	Ticks, *Boophilus microplus, B. decolaratus, B. annulatus, B. geigyi, Rhipicephalus evertsi*	Central and South America, Southern Europe, Africa, Asia, Australia	Pathogenic: Bovine babesiosis (redwater fever or tick fever or Texas fever or piroplasmosis or ranilla, or splenic or La Tristeza) in cattle
35.	Protozoan, *Babesia bovis* (Sporozoea: Babesiidae)	Cattle, man	Blood, spleen	Ticks, *Ixodes* spp., *Rhipicephalus* spp., *Boophilus microplus, Boophilus annulatus, Boophilus geigyei*	Central and South America, Southern Europe, Africa, Asia, Australia	Pathogenic (& zoonotic): Bovine babesiosis in cattle and human babesiosis
36.	Protozoan, *Babesia major* (Sporozoea: Babesiidae)	Cattle	Blood, spleen	Ticks, *Haemophysalis punctata.*	Europe, North Africa	Pathogenic: Bovine babesiosis in cattle
37.	Protozoan, *Babesia jakimovi* (Sporozoea: Babesiidae)	Cattle	Blood, spleen	Ticks, *Ixodes ricinus*	Siberia, Northern Russia	Pathogenic: Bovine babesiosis in cattle
38.	Protozoan, *Babesia ovata* (Sporozoea: Babesiidae)	Cattle	Blood, spleen	Ticks, *Haemophysalis longicornis.*	Japan	Pathogenic: Bovine babesiosis in cattle
39.	Protozoan, *Babesia occultans* (Sporozoea: Babesiidae)	Cattle	Blood, spleen	Ticks, *Hyalomma m. rufipes*	South Africa	Pathogenic: Bovine babesiosis in cattle
40.	Protozoan, *Babesia divergens* (Sporozoea: Babesiidae)	Cattle, man	Blood, spleen	Ticks, *Ixodes ricinus, I. persulcatus, Haemophysalis* sp.	Northern Europe	Pathogenic (& zoonotic): Bovine babesiosis in cattle
41.	Protozoan, *Babesia microti* (Sporozoea: Babesiidae)	Man, rodents, insectivores	Blood, spleen	Ticks, *Ixodes dammini*	USA, Europe	Pathogenic (& zoonotic): Human babesiosis
42.	Protozoan, *Babesia caballi* (Sporozoea: Babesiidae)	Horse, mule, donkey	Blood	Ticks, *Dermacentor* spp., *Hyalomma* spp., *Rhipicephalus* sp.	World-wide	Pathogenic: Equine babesiosis
43.	Protozoan, *Babesia equi* (Sporozoea: Babesiidae)	Horse, mule, donkey	Blood	Ticks, *Dermacentor* spp., *Hyalomma* spp.,	World-wide	Pathogenic: Equine babesiosis (biliary fever) in horses

				Rhipicephalus spp.		
44.	Protozoan, *Babesia canis* (Sporozoea: Babesiidae)	Dog	Blood	Ticks, *Rhipicephalus* spp., *Dermacentor* spp., *Haemophysalis* spp., *Hyalomma* spp.	Southern Europe, North America, Africa, Asia	Pathogenic: Canine babesiosis (tick fever) in dogs
45.	Protozoan, *Babesia gibsoni* (Sporozoea: Babesiidae)	Dog	Blood	Ticks, *Rhipicephalus sanguineus*, *Haemophysalis bispinosa*	Far East, Asia, Europe, North America	Pathogenic: Canine babesiosis in dogs
46.	Protozoan, *Babesia motasi* (Sporozoea: Babesiidae)	Sheep, goats	Blood	Ticks, *Dermacentor* spp., *Haemophysalis* spp., *Hyalomma* spp., *Rhipicephalus* spp.	Europe, Asia, Africa	Pathogenic: Ovine/caprine babesiosis in sheep and goats
47.	Protozoan, *Babesia ovis* (Sporozoea: Babesiidae)	Sheep, goats	Blood	Ticks, *Rhipicephalus* spp., *Ixodes* spp.	Southern Europe, Asia (Middle East), Africa	Pathogenic: Ovine/caprine babesiosis in sheep and goats
48.	Protozoan, *Babesia trautmanni* (Sporozoea: Babesiidae)	Pigs	Blood	Ticks, *Boophilus* spp., *Rhipicephalus* spp., *Hyalomma* spp., *Dermacentor* spp.	Southern Europe, former USSR, Asia, Africa	Pathogenic: Ovine/caprine babesiosis in pigs
49.	Protozoan, *Babesia perroncitoi* (Sporozoea: Babesiidae)	Pigs	Blood	Unknown	Southern Europe, Africa (Sudan), Sardinia	Pathogenic: Ovine/caprine babesiosis in pigs
50.	Protozoan, *Babesia felis* (Sporozoea: Babesiidae)	Cats	Blood	Unknown	Africa, Southern Asia	Pathogenic: Feline babesiosis in cats
51.	Protozoan, *Babesia herpailuri* (Sporozoea: Babesiidae)	Cats	Blood	Unknown	Africa	Pathogenic: Feline babesiosis in cats
52.	Protozoan, *Theileria parva* (Sporozoea: Theileriidae)	Cattle, African buffalo	Blood, lymph glands, lymphatic system, endothelial cells	Ticks, *Rhipicephalus appendiculatus*	Eastern and central tropical Africa: Kenya, Uganda, Tanzania, Malawi, Zambia,	Pathogenic: East Coast Fever (ECF), severe in cattle

					Mozambique, Rhodesia, Rwanda, Burundi, Congo, Sudan	
53.	Protozoan, *Theileria annulata* (Sporozoea: Theileriidae)	Cattle, water buffalo	Blood, lymph glands, lymphatic system, endothelial cells	Ticks, *Hyalomma excavatum*, *Hyalomma detritum*	Southern Europe, North Africa, Central Asia (Iran)	Pathogenic: Tropical Theileriosis (Mediterranean Coast Fever-MCF), severe in cattle
54.	Protozoans, *Theileria bovis/T. lawrencei* (Sporozoea: Theileriidae)	Cattle	Blood, lymph glands, lymphatic system, endothelial cells	Ticks, *Rhipicephalus* spp., *Hyalomma* spp.	South Africa	Pathogenic: Corridor Diseases (Zimbabwean Theilerioses), severe in cattle
55.	Protozoan, *Theileria mutans* (Sporozoea: Theileriidae)	Cattle	Blood, lymph glands, lymphatic system, endothelial cells	Ticks, *Rhipicephalus* spp., *Amblyomma* spp.	Widespread	Low pathogenicity
56.	Protozoan, *Theileria ovis* (Sporozoea: Theileriidae)	Sheep, goats	Blood, lymph glands, lymphatic system, endothelial cells	Ticks, *Rhipicephalus* spp., *Amblyomma* spp.	Widespread	Low pathogenicity
57.	Protozoan, *Theileria hirci* (Sporozoea: Theileriidae)	Sheep, goats	Blood, lymph glands, lymphatic system, endothelial cells	Unknown	North and East Africa, Southern Europe	Pathogenic: Malignant Theileriosis, severe in sheep called ovine theileiosis
58.	Protozoan, *Anaplasma marginale*	Cattle	Blood	Ticks, *Boophilus* spp., *Dermacentor* spp., *Ixodes* spp., *Hyalomma* spp., tabanid flies (deer flies, horse flies)	Latin America, Southern Europe, North Europe, Africa, Asia, Australia	Pathogenic: Anaplasmosis (tick-borne gall sickness) in cattle
59.	Protozoan, *Anaplasma centrale*	Cattle	Blood	Ticks, *Boophilus* spp., *Rhipicephalus* spp., *Dermacentor* spp., *Ixodes* spp., *Hyalomma* spp., tabanid flies (deer flies, horse flies) e.g. *Stomoxys* spp.	World-wide	Pathogenic: Anaplasmosis in cattle
60.	Protozoan,	sheep	Blood	Ticks, *Boophilus*	World-wide	Pathogenic:

	Anaplasma ovis				spp., *Dermacentor* spp., *Ixodes* spp., *Hyalomma* spp., tabanid flies (deer flies, horse flies)		Anaplasmosis in sheep
61.	Protozoan, *Aegyptianella pullorum*	Fowl	Blood	Ticks, *Argaspersicus*	Africa, Asia, Southeastern Europe	Pathogenic: Anaplasmosis in fo
62.	Protozoan, *Eperythrozoon wenyoni*	Cattle	Blood	Lice, mechanical transmission by haematophagous flies	Widespread	Non-pathogenic
63.	Protozoan, *Eperythrozoon ovis*	Sheep, goats	Blood	Lice, mechanical transmission by haematophagous flies	Widespread	Non-pathogenic
64.	Protozoan, *Eperythrozoon parvum*	Pigs	Blood	Lice, mechanical transmission by haematophagous flies	Widespread	Non-pathogenic
65.	Protozoan, *Eperythrozoon suis*	Pigs	Blood	Lice, mechanical transmission by haematophagous flies	Widespread	Low pathogenicity pigs
66.	Protozoan, *Haemobartonella* spp.	Cattle, goats, pigs, dogs, cats	Blood	Lice, mechanical transmission by haematophagous flies	Widespread	Non-pathogenic
(b) Filarial nematode parasites						
67.	Filarial nematode, *Litomosoides carinii* (Spirurida: Filarioidea)	Cotton rat, *Sigmodon hispidu;,* squirrels, *Sciurus;* rats, *Neotoma; Mus;* and, *Holochilus*	Pleural cavity, peritoneal space, lung, muscle fibres, cardiac muscle fibres of the ventricles	Mite, *Ornithonyssus bacoti*	USA, South America	Not reported
68.	Filarial nematode, *Dirofilaria immitis* (Spirurida: Filarioidea)	Man, dogs, cats, wolves, foxes and other carnivores	Blood, heart	Mosquito: *Culex* spp. (Diptera: Culicidae)	World-wide	Pathogenic (& zoonotic): Dirofilariasis in do
69.	Filarial nematode, *Wuchereria bancrofti* (Spirurida: Filarioidea)	Man	The lymph nodes and lymphatics	Mosquito: members of *Culex (C.) pipiens* complex in urban areas and species of *Anopheles*, *Aedes* and more rarely *Mansonia*.	Tropical and subtropical countries; in Asia, Africa, America and the Pacific	Pathogenic: Bancroftian filaria (elephantiasis) in man
70.	Filarial nematode,	Man,	Lymph nodes	Mosquito (	South East	Pathogenic (&

	Brugia malayi (Spirurida: Filarioidea)	monkeys, wild cats, pangolin	and lymphatics	Culicidae), *Mansonia* spp. and *Anopheles* spp.	Asia: India, Sri Lanka, Thailand, Malaysia, Korea, Japan, Philippines	zoonotic): Malayan filariasis in man
71.	Filarial nematode, *Brugia timori* (Spirurida: Filarioidea)	Man	Lymph nodes and lymphatics	Mosquito (Culicidae)	Indonesia	Pathogenic
72.	Filarial nematode, *Brugia pahangi* (Spirurida: Filarioidea)	Cats and dogs	Lymph nodes and lymphatics	Mosquito (Culicidae)	Malaysia, Borneo	Not reported
73.	Filarial nematode, *Loa loa* (Spirurida: Filarioidea)	Man, monkeys	Subcutaneous tissue, eye	Tabanid flies, *Chrysops silacea*, *Chrysops diminiata*	West and Central Africa	Pathogenic (& zoonotic): Eye worm disease in man
74.	Filarial nematode, *Mansonella perstans* (Spirurida: Filarioidea)	Man, apes	Body cavities, tissues	Midges, *Culicoides* spp.	Africa, South America	Pathogenic (& zoonotic): Mild in man and apes
75.	Filarial nematode, *Mansonella azzardi* (Spirurida: Filarioidea)	Man, horses, cattle, antelopes	Peritoneal cavity	Midges, *Culicoides furens*	West Indies, Northern parts of South America	Non-pathogenic
76.	Filarial nematode, *Mansonella streptocera* (Spirurida: Filarioidea)	Man, chimpanzees	Peritoneal cavity	Midges, *Culicoides grahamii*, other *Culicoides* spp.	Ghana, Cameroon, Zaire	Not reported
77.	Round worm (filarial nematode), *Onchocerca volvulus* (Spirurida: Filarioidea)	Man, domestic animals (cattle)	Blood, under the skin, between muscles, ligaments tendons; and in nodules (onchodermata) under the skin	Black fly, *Simulium* spp. (Diptera: Simuliidae)	East, central and western Africa; Asia; South Pacific; and North America	Pathogenic (& zoonotic): Human onchocerciasis (river blindness), cattle onchocerciasis
78.	Round worm (filarial nematode), guinea worm, *Dracunculus medinensis* (Spirurida: Filarioidea)	Man	Subcutaneous tissues; ankle and foot, arms and shoulders, and the whole body	Copepod, *Cyclops* sp.	Tropical Africa, India, Pakistan, Saudi Arabia and Yemen	Pathogenic: Dracunculosis in man
(c) Trematode parasites						
79.	Trematode, *Paragonimus westermanni (Trematoda: Troglotrematidae*	Man, carnivores (cats, dogs)	Lungs, brain, liver, spleen, intestinal wall, eye, muscles, kidneys	Crustaceans; crabs, crayfish	Korea, Japan, Taiwan, Central China, Philippines	Pathogenic (&zoonotic): Paragonimiasis in man, cats and dogs
80.	Trematode, species,	Man	Blood stream	Molluscan, snails,	Most of Africa,	Pathogenic:

	Schistosoma mansoni and others (Schistosomatidae).			*Biomphalaria* spp.	Egypt, Israel, some parts of Arab world except Iraq, Caribbean Islands, Southern USA, South America.	schistosomiasis (bilharzias) in man
(d) Cestode parasites						
81.	Cestode, fish tape worm of man or human broad tapeworm, *Diphyllobothrium latum* (Pseudophyllidea: Diphyllobothriidae)	Fish-eating mammals, including, dog, cats, man	Small intestine	Fresh water copepods, *Cyclops* sp.,	World-wide: especially in Europe in countries bordering the Baltic Sea, Finland, Sweden, etc; Russia, Switzerland and North America	Pathogenic (& zoonotic): Diphyllobothriasis (fish tapeworm infection) in man, pigs, dogs
82.	Cestodes, tape worms, *Diphylidium caninum* (Cyclophyllidea: Hymenolepididae)	Dogs, cats, foxes, man (children)	Intestine	Fleas: dog flea, *Ctenocephalides canis*; cat flea, *C. felis*; human flea, *Pulex irritans*; dog louse, *Trichodectes canis*	World-wide	Pathogenic (& zoonotic): Human dipylidiasis
83.	Cestodes, tape worms, *Diphyllobothrium* and *Spirometra* species (Pseudophyllidea: Diphyllobothriidae)	Frogs, snakes, hogs, hedgehogs, dogs, cats, man	Viscera	Fresh water copepods, *Cyclops* sp.,	Asia	Pathogenic (& zoonotic): Sparganosis (sparganum infection) in man, cats, dogs
84.	Cestodes, tape worm, *Hymenolepsis diminuta* (Cyclophyllidea: Hymenolepididae)	Rats, other rodents; other mammals, including man	Intestine	Flea, *Xenopsylla cheopis* and other various rodent fleas, *Ceratophyllus fasciatus*, *Ctenocephalides canis*, etc; and beetles, *Tenebrio molitor* (larva), *Tribolium confusum* (larva)	Cosmopolitan	Pathogenic (& zoonotic): Rat tapeworm infection
85.	Cestodes, tape worm, dwarf tapeworm, *Hymenolepsis nana* (Cyclophyllidea: Hymenolepididae)	Rodents and man	Intestine	Beetle, flour beetle	Cosmopolitan	Pathogenic (& zoonotic): Hymenolepiasis in m

Sources: Adam, *et al.* (1979); Smyth (1996); Okello-Onen, *et al.* (1999); Anonymous (2008); and Biryomumaisho, *et al.* (2012).

The vector range of other arthropod borne pathogens and diseases

This part outlines major arthropod-borne diseases caused by fungal, bacterial, rickettsial, viral pathogens; other than parasites. For convenience, however, they have been put into 2 groups: bacterial/rickettsial and viral, based on the findings in this work. From the summary in Table IV below, it is easy to follow a pathogen, its vector and the disease it causes. Your attention is being drawn more especially towards the vector which may be easier to follow and manage in case natural calamities expected out of global warming and climate change. It is possible from here to trace in future which pathogen has expanded its vector range.

Table: IV: A summary of the vector range of other arthropod borne pathogens and diseases

S/No.	Pathogen	Arthropod vector	Disease
(a) Bacterial/rickettsial pathogens			
1.	Gram negative bacterium, *Rickettsia rickettsia*	Ticks, *Dermacentor* sp. and other Ixodid ticks	Rocky Mountain Spotted Fever in man
2.	Bacterium (Rickettsia), *Rickettsia prowazekii*	Louse, *Pediculus humanus* (Anoplura: Pediculidae)	Epidemic typhus in man
3.	Spiral shaped bacterium, *Borrelia* sp.	Soft tick, *Ornithodoros* sp.	Endemic Relapsing Fever in man
4.	Spiral shaped bacterium, *Borrelia burgdorferi*	Tick, *Ixodes* sp.	Lyme disease in man
5.	Bacterium: *Ehrlichia canis, E. sennetsu, E. chaffeensis, E. equi, E. phagocytophilia*	Ticks, *Dermacentor variabilis, Amblyomma americanum*	Ehrlichiosis (sennetsu fever) in man
6.	Spiral bacterium, *Borrelia theileri*	Ticks, *Boophilus spp., Rhipicephalus evertsi evertsi*	Spirochaetosis in cattle, horses, sheep and goats
7.	Gram negative bacterium (Rickettsia), *Rochalimaea (Bartonella) quintana*	Louse, *Pediculus humanus* (Anoplura: Pediculidae)	Trench fever (bacillary angiomatosis or bacillary peliosis) in man
8.	Bacterium (spirochete), *Borrelia recurrentis*	Louse, *Pediculus humanus* (Anoplura: Pediculidae)	Louse-borne relapsing fever (epidemic relapsing fever) in man
9.	*Coxiella burneti*	Ticks, *Hyalomma* spp.	Q fever in man
10.	*Rickettsia conorii*	Ticks, *Hyalomma* spp.	Tick typhus in man
11.	*Pasturella pestis*	Ticks, *Hyalomma* spp.	Plague in man
12.	Bacterium, *Rickettsia akari*	Mouse mite, *Liponyssiodes sanguineus*	Rickettsial pox in man
13.	Bacterium, *Bacillus anthracis*	Mechanical transmission by horseflies and deerflies (Tabanidae)	Human anthrax
14.	Rod shaped gram negative bacterium, *Francisella tularensis*	Tabanid fly, *Chrysops* sp.; tick, *Dermacentor* sp.	Tularemia in man
15.	Intracellular bacterium, *Rickettsia tsutsugamushi*	Red mite, *Leptotrombidium sp.*	Scrub Typhus (Tsutsugamushi disease) in man
16.	Rod shaped gram negative bacterium, *Yersina pestis*	Flea, *Xenopsylla cheopis,* and various other rodent	Plague

		fleas	
17.	Bacterium, *Rickettsia typhi*	Flea, *Xenopsylla cheopis*	Murine typhus
18.	Gram negative bacterium, *Bartonella bacilliformis*	Sand fly, *Phlebotomus* sp.	Bartonellosis (Oroya fever or Carrion disease)
(b) Viral pathogens			
19.	Colorado Tick Fever (CTF) Virus	Tick, *Dermacentor* sp.	Colorado Tick Fever
20.	Sand fly fever Naples Virus or Sand fly Sicilian Virus or Rift Valley fever Virus (RVFV) (Bunyaviridae)	Sand fly	Sand fly fever (Rift Valley Fever)
21.	Venezualan Equine Encephalitis virus or Western Equine Enchephalitis virus (WEEV) (Togaviridae)	Mosquito: *Culex* sp., *Culex tarsalis* (Diptera: Culicidae)	Venezualan equine encephalitis (Western equine encephalitis) in ma:
22.	St. Louis Encephalitis virus (Flaviviridae)	Mosquito: *Culex* sp., *Culex pipiens* (Diptera: Culicidae)	St. Louis Encephalitis in man
23.	Russian Spring-Summer Encephalitis Virus or Louping III Encephalitis Virus or Langat Encephalitis Virus or Powassan Encephalitis Virus or Omsk haemorrhagic fever virus (Flaviviridae)	Tick	Russian Spring-Summer Encephali: (Louping III Encephalitis, Langat Encephalitis, Powassan Encephaliti· Omsk haemorrhagic fever) in man
24.	Nairobi sheep disease virus or Crimean-Congo haemorrhagic fever virus (Bunyaviridae)	Tick	Nairobi sheep disease virus or Crimean-Congo haemorrhagic fever in man
25.	Yellow fever virus (YFV) (Flaviviridae)	Mosquito: *Aedes aegypti*	Yellow fever in animals and man
26.	Eastern Equine Encephalitis Virus (EEEV) (Togaviridae)	Mosquito: *Culiseta melanura, Coquillettidia pertubans, Aedes vexans*	Eastern Equine Encephalitis
27.	La Crosse Encephalitis Virus (LACV) (Bunyaviridae)	Mosquito: *Aedes triseriatus*	La Crosse Encephalitis
28.	Chikungunya virus (Mayaro fever virus, Mucambo fever virus, O'Nyong-Nyong fever virus, Pixuna fever virus or Ross River fever virus) (Togaviridae)	Mosquito	Chikungunya forest fever
29.	Nile fever virus (Japanese encephalitis virus, West Nile Virus, Zika fever virus, Wesselsbron fever virus, or Kyasanur forest disease virus) (Flaviviridae)	Mosquito	Fevers and encephalitis
30.	Oropouche virus (Bunyamwera virus, Bwamba fever virus, Guama fever virus, Oropouche fever virus, or California Encephalitis virus) (Bunyaviridae)	Mosquito	Fevers and encephalitis
31.	Chandipura fever virus (Piry fever virus) (Rhabdoviridae)	Mosquito	Fevers
32.	Dengue fever virus (DeV) (Flaviviridae)	Mosquito: *Aedes* sp.	Dengue fever (Break bone fever) in man

Sources: Buxton (1955); Hamilton (2013); and Anonymous (2013).

Discussions

We have seen the established host, habitat and geographical distribution pattern of arthropod-borne parasites (Tables II and 3). This pattern was established over years of researches done when environmental conditions were still average. However, it is apparent that something has changed and is still changing in the environment. The best testimony to this change comes from a rural setting where agriculture is the main means of livelihoods. Farmers used to be clear of the rainfall pattern in their areas and would therefore plan when to dig, plant and harvest for the various crops in a year's time. Today, the rainfall pattern is distorted; farmers can no longer predict rain correctly and Government Departments responsible for weather and climate forecasting are being blamed for not telling the truth. We are aware that under natural conditions, the earth is a self-regulating (homeostatic) unit, balancing all factors of the environment, including temperature. Earth's atmosphere allows heat from the sun in, but only lets a certain amount out, which is a result of an interplay between ozone layer and certain gases in the atmosphere, so that the earth can support life like a green house (Environmental Alert, 2010). As such, these gases are called greenhouse gases. Therefore, greenhouse gases are gaseous elements of the atmosphere that absorb and emit radiation from the sun to keep the earth warm and habitable for all living things on it. Examples of greenhouse gases are carbon dioxide, methane, and nitrous oxide. The ozone layer acts like a blanket to protect the earth from direct heat from the sun. It has also been established that greenhouse gases react with ozone layer, and under natural conditions there is a balance. However, excess greenhouse gases increase the rate of reaction and depletion of the ozone layer. When the thin ozone layer is depleted, sun rays hit directly on the earth resulting in rise in temperature. Average global temperatures are on the rise, making the 20th century the warmest, the world has ever seen in 1,000 years since the 1980s were the warmest decades on record. This rise in temperature of the globe is a phenomenon now called global warming and is accompanied by climate change and its effects. Global warming is therefore caused by the production of excess greenhouse gases from human activities, including agricultural production, industrialization, burning of fossil and bio fuels, and deforestation among others (Environmental Alert, 2010). High global temperatures disorganize the balance set in environmental conditions, leading sometimes to extremes such as floods, desertification, and epidemic outbreaks of diseases. According to Environmental Alert (2010), climate change alters precipitation (rainfall) patterns, water availability, length of seasons, among others, including distribution and prevalence patterns of pests and diseases. There is apparent alteration of riverine ecosystems, habitat change disruption for riverine arthropod vectors like mosquitoes and tsetse flies. There will be rampant pest and disease outbreaks as a result of increasing mosquito and tsetse fly spread. As a result of the ever increasing human population, forests and swamps will continue to be encroached on for agriculture, industries, and settlement, thereby further increasing human-vector and animal-vector contact. Nomadism is set to increase in search of water and pasture by pastoral communities, and the speed of international travels will also increase in relation to world population growth and globalization. Increasing cross-border interactions are part of the effects of global warming and climate change. Reversing global warming may be far from reality, but one immediate option is climate change adaptation strategy. Man is now in search of drought resilient crop and animal species. Parasites will probably adapt to this change by expanding their host, habitat and geographical distribution range. Based on the explanations above, interactions among domestic animals, man, wild animals, arthropod vectors, parasites and pathogens is on the rise and crossing borders. All parasites should be considered potentially pathogenic as they have ever proved so in situations of lowered immunity of the host victim. We are aware that global warming and climate are stressing to organisms, and impair their immunity. All hosts should be considered the same; concepts of intermediate and definitive hosts will fail to be operational as parasites may change their mode of life to survive in the changing circumstances. Although this review has not captured all the information that is required, it has contributed the basic that is required to build on in future for tracing global warming and climate change effects, a move in the direction to one world one health.

General Recommendations

Each country should research, document and/or review arthropod-borne parasites, pathogens and diseases so as to be able to trace their whereabouts in future in face of global warming and climate change.

Host, habitat and Geographical Range strategy to management of arthropod-borne parasites should be adapted by planners and implementers of human and animal health projects and programmes, as it offers a more protracted approach to tackling the problems.

Most arthropod-borne diseases are caused by protozoan parasites; more attention should therefore be put on protozoology.

Pathogenicity of parasites with narrow host ranges like *Anaplasma* and A*naplasma*-related spp., *Theileria* spp., *Leishmania* spp., and trematodes should be tested in other animals, to assess degree of virulence before hand.

More stringent local and international laws, rules and standards are required to govern movement of animals and animal products, in order to control spread of arthropod-borne diseases.

The ordinary tax payer is becoming more and more desperate each day with the effects of global warming and climate change, and the responsibility of state and central governments in the health of their people is correspondingly adding, which should not be continuously ignored.

Conclusion

The level of documentation of arthropod-borne parasites and diseases varies from country to country and continent to continent with most of the information being generalized at continental level. This is likely to hamper implementation and outcome of the host, habitat and geographical range strategy, which requires precise accuracy.

Protozoans of *Trypanosoma* spp. have a wide range of hosts. This capacity can make them expand their host range more easily than any other parasite.

Filarial nematode worms also rank high in the range of hosts. They are also likely to increase their host range more easily.

Ticks surpass all other arthropods in the number and variety of diseases they transmit to animals and man. Their elimination, eradication, or suppression will accordingly control the diseases they transmit.

Mosquitoes also rank high in the number and variety of diseases they transmit, their elimination, eradication, or suppression will accordingly control mosquito-borne diseases.

Bibliography

- ADAM, KATHERINE M.G.; PAUL, JAMES; and ZAMAN, VIQAR (1979). Medical and Veterinary Protozoology. An illustrated guide. Longman Group Limited. Churchill Livingstone, Edinburgh and London. Great Britain.
- ANONYMOUS (2012a). Insect orders (Online). Entomology at Texas A and M University. Available at www.insects.tamu.edu/fieldguide/orders.htm (Retrieved 29 November 2012).
- ANONYMOUS (2013). Arthropod-Borne Diseases (Online). Available at http://www.atsu/faculty/chmberlain/arthro.htm (Retrieved 14 January 2013).
- ANONYMOUS. Short Certificate Course Training in Diagnostics and Management of Ticks and Tick-Borne Diseases (TBDs). Makerere University Faculty of Veterinary Medicine. Kampala. Uganda. 2008: p56.
- BIRYOMUMAISHO, S; MUNYAGISHARI, E; INGABIRE, D; and GAHAKWA, D. "Risk factors that influence the Distribution and Acaricide Susceptibility of Ixodid ticks infesting cattle in Rwanda". African Union Interafrican Bureau for Animal Resources (AU-IBAR) Bulletin of Animal Health and Production in Africa. Vol. 60 No. 2. June, 2012. Nairobi Kenya. 2012:p139-147.
- BIRYOMUMAISHO, SAVINO. "Control of Trypanosomiasis in Animals". A short Certificate Course in New Techniques in Tsetse and Trypanosomiasis Control. Makerere University Faculty of Veterinary Medicine. Kampala. Uganda. 2007b: p51-56.
- BIRYOMUMAISHO, SAVINO. "Potential and Practice of Community Participation in Tsetse and Trypanosomiasis Control". A short Certificate Course in New

Techniques in Tsetse and Trypanosomiasis Control. Makerere University Faculty of Veterinary Medicine. Kampala. Uganda. 2007c: p80-83.

- BIRYOMUMAISHO, SAVINO. "Trypanosomiasis: A disease of Animals and People". A short Certificate Course in New Techniques in Tsetse and Trypanosomiasis Control. Makerere University Faculty of Veterinary Medicine. Kampala. Uganda. 2007a: p43-49.
- BUXTON, P.A. "Karl Jordan's Contribution to our Knowledge of Fleas associated with Disease." A collection of essays and scientific papers brought to celebrate the ninety-fourth birth day of Karl Jordan. The transactions of the Royal Entomological Society of London. vol. 107. 6th December 1955. 1955:p43-44.
- CAMBRIDGE UNIVERSITY PRESS (1995). Cambridge International Dictionary of English. Cambridge.
- CDC. Malaria Worldwide Impact. (Online). Available at http://www.cdc.gov/malaria/malaria_worldwideimpact.html (Retrieved 28 December 2012a)
- CDC. Sleeping Sickness. (Online). Available at http://www.cdc.gov/parasites/sleepingsickness/ (Retrieved 28 December 2012b)
- DAVIES, R. G. (1988). Outlines of Entomology. 7th edition. Chapman and Hall. London.
- DRANSFIELD, R.D. (1988). "Ecological and Epidemiological Data base required for Control of Animal and Human Trypanosomiasis". Integrated Tsetse and Trypanosomiasis Control in Uganda. Proceedings of a workshop held in Mukono, Uganda, 20-23 May 1987. The Republic of Uganda.
- ENVIRONMENTAL ALERT (2010). Climate Change in Uganda. Insights for Long term Adaptation and Building Community Resilience. An Issues Paper. July 2010.
- FAO (1982a). Training Manual for Tsetse Control Personnel. Vol. 1: Tsetse Biology, Systematics and Distribution Techniques. FAO. Rome. 279pp.
- FAO (1982b). Training Manual for Tsetse Control Personnel. Vol. 2: Ecology and Distribution of Tsetse. FAO. Rome. 101pp.
- FAO (1982c). Training Manual for Tsetse Control Personnel. Vol. 3: Control Methods and Side Effects. FAO. Rome. 128pp.
- FAO (1984a). Ticks and Tick-borne Diseases Control. A Practical Field Manual. Volume I: Tick Control. Food and Agriculture Organization of United Nations. Rome.
- FAO (1984b). Ticks and Tick-borne Diseases Control. A Practical Field Manual. Volume II: Tick-Borne Disease Control. Food and Agriculture Organization of United Nations. Rome.
- FAO (1992). Training Manual for Tsetse Control Personnel. Vol. 4: Use of Attractive Devices for Tsetse Survey and Control. FAO. Rome. 196pp.
- FAO (1993). Training Manual for Tsetse and Trypanosomiasis Control, Using Attractive Bait Techniques. FAO. Rome. 88pp.
- FORD, J. (1968). "The Control of populations through limitation of Habitat Distributions as exemplified by Tsetse flies". Insect Abundance. Southwood, T.R.E (ed.). Symposia of the Royal Entomological Society, London.

- GHAFFAR, ABDUL; AND HUNT, RICHARD. "Parasitology. Arthropods". Microbiology and Immunology. University of South Carolina School of Medicine. (Online). Available at http://pathmicro.med.sc.edu/book/para-sta.htm (Retrieved 14 January 2013).
- HAMILTON, JO. Parasitic Arthropods II-Power Point PPT Presentation. (Online). Available at http://www.powershow.com/view/54e2d-OTg4N/presentation (Retrieved 14 January 2013).
- HARRIS, J. PAT. "Venomous Insects". Grounds Maintenance (Online). Available at http://grounds-mag.com/mag/grounds_maintenance_venomous_arthropods/ (Retrieved 14 January 2013).
- JORDAN, A.M. (1961). An Assessment of the economic importance of tsetse species of Southern Nigeria and Southern Cameroon based on their trypanosome infection rates and ecology. Commonwealth Institute of Entomology, London.
- JORDAN, A.M. (1986). Trypanosomiasis Control and African Rural Development. Longman Group Limited, London. 357pp
- KALYEBI, A. (1998). The Tsetse flies (Diptera: Glossinidae) of Buvuma Island, Lake Victoria, Uganda. M.Sc. Thesis, Makerere University, Kampala.
- KANGWAGYE, T.N. (1968). A study of interactions between biting flies and large mammals in Western Uganda. PhD Thesis. Makerere University, Kampala.
- KANGWAGYE, T.N. (1988a). "The distribution of Tsetse flies". Integrated Tsetse and Trypanosomiasis Control in Uganda. Proceedings of a workshop held in Mukono, Uganda, 20-23 May 1987. The Republic of Uganda.
- KANGWAGYE, T.N. (1988b). "The Tsetse fly Problem". Integrated Tsetse and Trypanosomiasis Control in Uganda. Proceedings of a workshop held in Mukono, Uganda, 20-23 May 1987. The Republic of Uganda.
- LOCKE, M; and SMITH, D.S. (eds). (1980). Insect Biology in the Future. Academic Press, New York.
- MUGASA, CLAIRE. "Bait Technology for Tsetse Control". A short Certificate Course in New Techniques in Tsetse and Trypanosomiasis Control. Makerere University Faculty of Veterinary Medicine. Kampala. Uganda. 2007b: p57-64.
- MUGASA, CLAIRE. "Tsetse Biology and Ecology". A short Certificate Course in New Techniques in Tsetse and Trypanosomiasis Control. Makerere University Faculty of Veterinary Medicine. Kampala. Uganda. 2007a: p5-36.
- NASH, T.A.M. (1969). Africa's Bane: The Tsetse fly. Collins. London.
- OKELLO-ONEN, JOSEPH; HASSAN, SHAWGI, M.; and ESSUMAN SULIMAN (1999). Taxonomy of African Ticks. An Identification Manual. ICIPE Science Press. Nairobi, Kenya.
- *Science* Daily. Lactating Tsetse Flies Models for Lactating Mammals? (Online). Available at http://www.sciencedaily.com/releases/2012/04/120418162302.htm (Retrieved 23 April 2012).
- SERVICE, M.W. (1980). A guide to Medical Entomology. Macmillan Press Ltd., London.
- SMART, J.; JORDAN, K. and WHITTICK, R. J. (1943). A handbook for the identification of insects of medical importance. British Museum. London.
- SMITH KLINE BEECHAM Chart of some important ecto-parasites

- SMYTH, J.D. (1996). <u>Animal Parasitology.</u> Cambridge Low Price Editions. Cambridge University Press.
- TURNER, D.A. (1980). "Tsetse ecological studies in Niger and Mozambique. 1. Population sampling". <u>Insect Sci. Applic.</u> 1, 9-13.
- TURNER, D.A. (1987). "The population ecology of *Glossina pallidipes* Austen (Diptera: Glossinidae) in the Lambwe Valley, Kenya. 1. Feeding Behaviour and Activity Patterns". <u>Bull. Ent. Res.</u> 77, 317-333.
- VREYSEN, M. J. B.; and KHAMIS, I.S. (1999). "Notes on the Ecology of a Natural *Glossina austeni* (Diptera: Glossinidae) Population in the Jozani Forest, Unguja Island of Zanzibar". <u>Insect Science and its Application.</u> ICIPE Science Press. Nairobi, Kenya. Vol. 19 No.2/3: p99-256.
- WAISWA, CHARLES. "Application of Insecticides used in Spraying". <u>A short Certificate Course in New Techniques in Tsetse and Trypanosomiasis Control.</u> Makerere University Faculty of Veterinary Medicine. Kampala. Uganda. 2007a: p65-67.
- WAISWA, CHARLES. "Commercialization of Tsetse and Trypanosomiasis Control". <u>A short Certificate Course in New Techniques in Tsetse and Trypanosomiasis Control.</u> Makerere University Faculty of Veterinary Medicine. Kampala. Uganda. 2007c: p73-77.
- WAISWA, CHARLES. "Districts Capacity to Control Tsetse and Trypanosomiasis". <u>A short Certificate Course in New Techniques in Tsetse and Trypanosomiasis Control.</u> Makerere University Faculty of Veterinary Medicine. Kampala. Uganda. 2007d: p78-79.
- WAISWA, CHARLES. "New Technologies in Control of Tsetse and Trypanosomiasis (SOS Approach)". <u>A short Certificate Course in New Techniques in Tsetse and Trypanosomiasis Control.</u> Makerere University Faculty of Veterinary Medicine. Kampala. Uganda. 2007b: p68-72.
- WHO (1987). <u>WHO Expert Committee on Onchocerciasis.</u> Third Report. Technical Report Series 752. Geneva. 167pp.
- WHO (1995). <u>Vector Control for Malaria and other Mosquito-borne Diseases.</u> Report of a WHO Study Group. Technical Report Series 857. Geneva. 91pp.
- WIKIPEDIA. <u>Entomology</u> (Online). Available at <u>http://en.wikipedia.org/wiki/Entomology</u> (Retrieved 19 December 2012b).
- WIKIPEDIA. <u>Taxonomy</u> (Online). Available at <u>http://en.wikipedia.org/wiki/Taxonomy</u> (Retrieved 7 November 2012a).
- WWH. <u>History of Entomology.</u> (Online). Available at <u>http://what-when-how.com/insects/history-of-entomology.insects/</u> (Retrieved 19 December 2012).
- YOUNG, A. (1982). <u>Population Biology of Tropical Insects.</u> Plenum Press, New York, USA.
- YU, P.; HABTEMARIAN, T.; ORYANG, D.; OBASA, M.; NGANWA, D.; and ROBNETT, V. (1996). "Stochastic model of tsetse flies (Diptera: Muscidae)". <u>Environmental Entomology.</u> 25 (1). 78-84. vol. 19, Part 2, 1996. Numbers 9398-9529.